# Contents

*Third Edition*

# SYSTEMS ANALYSIS FOR BUSINESS DATA PROCESSING

## by *H. D. Clifton*

*Principal Lecturer,*
*The Polytechnic, Wolverhampton*

BUSINESS BOOKS
COMMUNICA - EUROPA

*First published 1969*
*Second edition 1972*
*Second impression 1974*
*Third impression 1974*
*Third edition 1978*

ISBN 0 220 66369 6

This book has been set 11 on 12pt Press Roman.
Printed by Thomson Litho Ltd., East Kilbride, Scotland.
Bound by Mansell Bookbinders Limited, Witham, Essex.
For the publishers, Business Books Limited,
24 Highbury Crescent, London N5.

# *Preface*

During the past twenty years a growing awareness of the potentialities of data processing in business has lead to a dramatic upsurge in the numbers of computers used for commercial and industrial applications. This usage now permeates all aspects of business and government worldwide and brings with it an increasing demand for systems analysts at all levels. The minimum size of the organizations employing computer-based data systems has decreased to the point where no company can now afford to be without at least one systems specialist.

This book is intended for prospective systems analysts and data processing managers, and in particular for those with a few years of previous business experience. The contents also cover the needs of accountants and business managers who might wish to acquire a basic understanding of what data processing planning involves from their own viewpoints; for these people the more technical sections, such as Chapters 5 to 8, can be omitted. Students of business studies and computer science will find considerable material that is included in their college courses, and this is also true for persons attending short or part-time courses on systems analysis.

The text does not purport to be an introduction to the technicalities of computers; this subject is well covered by a variety of other publications. The reader who comes completely fresh to data processing is advised to acquire a basic knowledge of computing equipment before studying the chapters on files (Chapters 6 to 8). Readers with some data processing background, such as programmers, should find the entire contents to be at a suitable level.

This, third, edition contains a considerable number of references so as to enable readers to study certain topics in greater depth and also to provide general background reading based on both practical and theoretical ideas.

H.D.C.

# Chapter One

# *Introduction*

## 1.1 The meaning of data processing

The term 'data processing' has come into use to describe all business and administrative aspects of information and communication. Data can be regarded as facts that are known and from which inferences may be made, and, within the framework of data processing, appertains to the identification and measurements of objects, events and people, generally known as 'entities'. 'Processing' refers to any series of actions or measures that converts data into usable information. The implication is that data is not, in itself, usable, and this is true to the extent that it does not normally lead to direct action of any significance. By data being 'processed', however simply, information is extracted from which future actions and policies can be decided.

What then is so different about data processing that gives it the right to have the headlines? After all, data has been with us a long time, and man has been making inferences since his evolution as an intelligent thinker. Men also provided themselves with usable information long before the advent of electronic machines. When given a description of data processing, the clerk, company chairman, storeman, city treasurer, shopkeeper have justification for asking: 'What is special about data processing from my viewpoint?' These people, among many others, have long been involved with processed data in the course of their everyday jobs. The chairman presents his company's balance sheet only after a considerable amount of data has been processed. The shopkeeper checking his bills hardly sees himself in the role of data processor, but, for a time at least, that is his function.

We must avoid thinking of data processing as merely financial procedures; data can apply to the characteristics of events, persons and things as authentically as it does to money. Data subjected to processing may be drawn from many spheres of activity, and it

1

includes items such as road accident details, examinations marks, share prices and seat reservations. In the broad sense data processing involves all data from which useful information can result and, in the business world, this applies especially to information that enables management to make far-reaching decisions.

The essence of data processing lies in the ability of computers to digest vast amounts of raw data at high speed. Methods and techniques are now being employed that were inconceivable before the days of electronic computation, data transmission and magnetic storage. This new power is based not so much on the computer's ability to perform complicated calculations, as on its organizing abilities, i.e. its capacity to sort, store and compare numbers and names. The storage devices now used as part of a computer complex facilitate the rapid scanning of large files of records, with the result that more data can be taken into consideration in preparing management information and in controlling the organization's day-to-day activities.

The compactness and high processing speed of magnetically recorded files facilitates the centralization of records within the data processing department and as a consequence of this, integration of the organization's activities becomes possible. The ability to transmit data over long distances at high speed allows a large number of widespread users to gain rapid access to centralized files. The users are thus able to keep the files up-to-date and, at the same time, obtain their own informational requirements.

The creation of an integrated data processing system necessitates systems investigation and design of the highest order. This work is of higher level than 'organization and methods', involving not only a complete re-appraisal of the organization's methods for achieving its objectives, but also top management in re-assessing their information requirements for present and future control of the organization. Data processing systems are now being designed to cater for management requirements of a much more flexible nature and, when desirable, they can be arranged to allow managers to participate directly in obtaining selected information.


## 1.2 Evolution of data processing

Business organizations have had key-operated accounting machines of various types at their disposal for over half a century. These machines have been developed from simple mechanical adding devices into today's sophisticated electronic accounting machines and visible record computers. Their means of output are now not only printed documents but also paper tape and magnetically striped cards. The latter feature gives them the ability to record carried-forward data in a form that is easily re-input to the machine, and this, together with an internally stored program of instructions, provides a limited degree of automatic operation.

A parallel but quite separate development was that of punched card machines. Punched cards, originally designed for special purposes such as the analysis of census statistics, were in use for over forty years in business and government departments. Punched card machines were originally somewhat crude electromechanical devices, and in view of their large size, capable of doing surprisingly little in the way of business tasks. The mainstay of the system was the sorter; this machine, although very slow by modern sorting standards, was unsurpassed when it came to sorting large volumes of data. After sorting the cards, a tabulator was used to accumulate totals, and since it had no means of printing, it would stop at the end of each group of cards so that the totals could be copied manually from its registers.

When the printing tabulator appeared in the early 1930s, the first move had been made towards an automatic data processing machine. Never before had an accounting machine been able to run continuously without human control. Accompanying the printing tabulator were various card arranging and punching machines, such as the collator (interpolator), interpreter, reproducer and automatic gang punch. The gang punch could be attached to a tabulator to provide carried-forward output in the form of cards for direct input into the tabulator during the next week's run.

It was not until the mid-1950s that punched card equipment made use of electronics, at which time its range was increased by the introduction of electronic calculators. These were able to do multiplication, division and simple comparisons under the control of a wired program of instructions. Later, more advanced types of calculator were able to carry out quite complex calculations; so much so that some were employed to do scientific calculations in the fields of atomic research and aircraft design. The employment of punched card machines in business organizations was generally only for the straightforward aspects of payroll, costing, stock control and accounting, and then only in the larger organizations. These had sufficient volumes of data to justify the use of these machines in spite of their restricted capabilities.

What then are the disadvantages of these two streams of equipment as compared with modern computers? The key-operated machines are geared inherently to the human operator. This has the advantage of versatility in that a wide range of tasks can be carried out with little time spent in preparation, but has the disadvantages of comparatively low speed of throughput and of human fallibility in operation. The punched card machines, on the other hand, after being set-up by the insertion of wired panels, were free from human error and were much faster in operation than the key-operated machines. They suffered, however, from the disadvantage of being inflexible; their functions were too limited to allow their use as management tools in the changeable situations that most businesses encounter.

A third and distinct development took place in the early 1950s in connection with scientific computing. Because of the tremendous increase in the complexity of scientifically designed plant and machines, such as atomic power stations, aircraft and weapon systems, the demand for computing power surged. This demand was largely met by the introduction of several models of computers, small by today's standards but ahead of their time in concept and design. These computers were mostly US developments but some of the early work was also done in the UK, mostly as a spin-off from war-time research in electronics and crytography. Unfortunately these machines had limited printing capacity and this, together with the absence at that time of high-speed mass-storage media, meant that in general, they were unsuitable as business computers, although a few were in fact used in this way.

The manufacturers of punched card machines also developed small computers based on the coupling of tabulators and gang punches to electronic units. These computers were typified by the IBM 650, several thousands of which were in use by the late 1950s. In common with the scientific computers, these punched card computers were controlled by internally stored instructions, and this feature gave them considerably more power than punched card machines for planning and analysis purposes. Nevertheless their speeds of input and output were geared to those of tabulators, and their only means of mass storage was punched card files. These limitations and the difficulty of programming made them unacceptable for most business applications. Another reason for their lack of success was the inability of both the business community and the manufacturers' salesmen to comprehend the potential of computers.

During the early 1960s IBM were able to win over many established punched card users on to the IBM 1400 series of computers. These were second-generation computers and had three main advantages over their predecessors: first, higher input/output speeds; second, a means of storing mass data and processing it at high speed, i.e. magnetic tape; third, higher-level languages to ease the programming burden.

The mid-1960s saw the introduction of the so-called third-generation computers, of which the world leaders in this field were the IBM system/360 and 370 series. The principal characteristics of third-generation computers that distinguished them from their forerunners were their extensibility and compatibility. Extensibility, as the name suggests, means that the user can make his computer grow with him by replacing or adding to its units (known as hardware). A small computer can thus be gradually transformed into a large one without at any one time replacing the computer as a whole, thereby minimizing the disturbance of changing over. Compatibility implies the ability of a given range of variously sized computers to accept each other's programs and a standard range of programs (software), also any selection from existing or future peripherals (hardware).

During the late 1960s the principal developments were in the field of peripheral equipment, arising from the compatibility aspect of third-generation computers. A considerable number of companies, some quite small, entered the data processing market with a wide range of peripheral units such as disc storage devices, visual display units and on-line input keyboards; also available is microfilm equipment for use with computers.

The main advance in the 1970s was the large-scale development of data transmission. This included the introduction of world-wide terminal-based systems operating in real-time mode, and the linking together of a number of computers to form computing networks.

Another significant development was the explosive increase in the software industry. Many small software houses came and went as the demand for software increased and the competition became severe. Those that survived have now attained a solid foundation, technically and financially.

Thus, although the mainframe suppliers have been reduced in number owing to take-overs and mergers, the over-all number of firms in the data processing industry has increased considerably.

An interesting account of the history and development of computers is given in [1.8].

## 1.3 Employment of computers in business

Computers are nowadays accepted as an intrinsic part of business and government organization. The amount of mundane but necessary work being done by computers is far greater than is generally realized and in most cases it would be impossible to return to manual methods. Clerical work, in the old-fashioned sense, is a dying occupation. When confronted with this fact, a natural but ludicrous response is, 'machines will never replace people'. How true — the introduction of data processing has resulted in an over-all increase in the demand for suitable staff.

The replacement of people has not been by machines but by different kinds of men—be they the original people. It is the nature of the occupation that has changed. Clerical staff now have at their disposal sophisticated computers of immense power to assist them in their tasks. The complexity of modern planning and the rate of change of many business situations are so great that they cannot really be coped with except by means of a man-computer relationship. Provided men can adjust their attitudes so as to assert their intelligence superiority over the computer, this relationship will be harmonious and rewarding.

*Business applications*

Computers are currently in use in very many companies and organizations; their application covers the gamut of conventional business and administrative activities. In addition they are assisting with a wide range of less well known applications, some of which at first glance may seem trivial or pointless. Nevertheless, if there is a demand for this work, and the most economic method is by employing a computer, these form the justification in themselves.

Several developments have helped in lowering the barriers between the 'ordinary' person and the computer; a brief mention of several developments in this direction is worthwhile at this point.

The bigger the computer, the better the value for money. It is therefore generally more economic to use a small amount of time on a big computer than a large amount of time on a small computer. The former aim can be achieved in one of two ways — either by renting time on a large computer and implementing all the work during that time, or by 'time-sharing' simultaneously with other users. Time-sharing involves the installation of terminal hardware and the provision of data transmission facilities, but it has a marked advantage over renting time in that the exact time of usage does not have to be pre-allocated, the user merely 'dials' the computer when he is ready and then makes immediate use of it.

Another aspect of man-computer communication has been the creation of common modes of communication. This is seen in an elementary form in the employment of stylized printing on documents; this is recognizable by machines and men, and is known as optical character recognition (OCR) or magnetic ink character recognition (MICR).

A more sophisticated form of common expression is the system known as 'conversational mode'. This brings the human and electronic counterparts into contact by permitting a stylized dialogue to take place between them. The computer is programmed to respond to the user in such a way as to make it appear that it understands his statements. At the same time the computer is able to lead the user along the correct path towards obtaining his requirements.

Allied with the developments mentioned above is the increased employment of visual methods and, to a lesser degree, audio input and output. The main visual technology is the visual display unit. This is arranged to provide a display of figures or graphs, and is able

6

to accept modifications to the data by means of a light-pen or similar device. The business application of this equipment has resulted in a much closer contact between the computer user and the files, the rapid speed of display enabling large amounts of data to be shown without delay thus facilitating the interrogation of files from a clerk's desk. Similarly the input of data, such as customers' orders, is brought closer to its source and is also moved nearer to the master files. These features provide for more stringent vetting of the input data and easier subsequent correction of the errors thus detected.

*Why analyse systems?*

The fragmented implementation of computer work can cause serious problems; it is dangerous to allow each department to specify its requirements in isolation, and then to design a system to provide these quite separately. The requirements of department A may conflict with those of department B; alternatively they may coincide, but in any event it is better that they are considered together. Only through a painstaking investigation and analysis of the over-all situation within the organization can a comprehensive plan evolve. This inevitably means that methodical systems analysis has to be carried out by competent staff working on a full-time basis.

   This is not to suggest that all applications must be implemented simultaneously, nor that it is absolutely imperative that every application is invariably included in the data processing system. The main objectives of systems analysis are (*a*) to study in depth the aims and problems of existing work, and (*b*) to design a system that is 'open-ended' so that further applications can be welded to it without duplication of work or records.

## 1.4   References and further reading

| | |
|---|---|
| 1.1 | MARTIN and NORMAN, *The computerised society*, Prentice Hall (1970). |
| 1.2 | 'Computers and social change',*Computers and Automation* (August 1970). |
| 1.31 | 'Technology and the future', *Data Systems* (December 1970). |
| 1.4 | 'Job satisfaction, *ibid*. (October 1970). |
| 1.5 | 'Computers and the ecology of management', *Data Processing* (July/August 1972). |
| 1.6 | ROTHERY, *The myth of the computer*, Business Books (1971). |

1.7    WILSON and MARTHANN, *What computers cannot do*, Auerbach (1971).

1.8    THIERAUF, Data processing for business and management, *Prologue* , John Wiley (1973).

1.9    'Computerisation, threat or promise?', *Data Systems* (March 1971).

1.10   'The human factor', *Data Processing* (July/August 1970).

1.11   MUMFORD and WARD, *Computers, planning for people*, Batsford (1972).

1.12   LAMBOURN, *Computer applications in business*, Longman (1974).

1.13   HIGGINS, *Information systems for planning control: concepts and cases*, Arnold (1976).

# Chapter Two
# *The systems analyst*

## 2.1   The need for full-time systems analysts

In the previous chapter the usages of hardware (machines and equipment) and software (the means of controlling machines and systems) have been mentioned briefly. We now turn to the employment of people within the orbit of data processing. Because of the considerable publicity given to computers and their electronic capabilities there has developed a tendency to overlook the people associated with them. Hard experience has shown that in many cases this has led to inefficient and fragmentary employment of expensive computing equipment.

Owing partly to this state of affairs, the occupation of 'Systems Analyst', as such, has evolved. His work was done previously by various other personnel, but within a narrower framework. These people were involved on a part-time basis and, more often than not, with isolated applications. Their efforts were frequently frustrated by lack of time and by insufficient training in data processing methods.

When the first computers were installed it was immediately recognized that programmers were necessary, and each user employed a few bright young men in this capacity. Early computer applications, such as payroll and sales analysis, were implemented successfully due to their inherent definability. Later applications, for instance production control, ran into difficulties owing to the programmers being unable to formulate them into neat computer routines. During the attempts to implement later applications it became apparent that a communication barrier existed between programmers and management. The former's attitudes were dominated by the intricacies of instructing the computer. Management, on the other hand, were adept at controlling men, but this skill had left them with a blindspot as regards the meticulous detail required for computer utilization.

This situation virtually halted the transference of further work on to the computer, so that it was used merely as a sophisticated accounting machine. The employment of a computer for integrated applications is not possible without a thorough investigation of the organization, and an over-all replanning of the existing systems.

These aims are best achieved through the employment of full-time systems analysts who are independent of the various departments in the organization. In some circumstances these staff work in a temporary capacity in project teams (Section 2.5), but during this phase they should be occupied solely with systems analysis work and so be totally relieved of their normal duties.

## 2.2 The function and duties of the systems analyst

Although the systems analyst can be regarded generally as an 'agent of change', his precise duties and responsibilities vary from one company to another. He is employed by organizations representing the whole spectrum of industrial, commercial and administrative society. His employers, in addition to the usual range of manufacturing and commercial companies, include consultancy firms, service bureaux, computer suppliers, educational institutions, and local and national government. It is therefore obvious that the range of his tasks is diverse; nevertheless the broad principles guiding his approach are the same for all situations.

The title 'Systems Analyst' is somewhat misleading and has caused the uninstructed to think in terms of physical systems such as power stations or chemical plants. The title has however now been assimilated into everyday language in connection with computers and data processing, and it is with this connotation that we are concerned here.

The systems analyst could equally well be entitled 'Systems Investigator', 'Systems Designer' or 'Systems Implementer' since all of these aspects are part of his function. His range of activities consists of :

1    The investigation and recording of existing systems with a view to discovering inefficiencies, problems and bottlenecks.
2    The analysis of data acquired during the investigation so as to prepare it for subsequent utilization in a data processing system.
3    The design and appraisal of new systems, usually involving computers, bearing in mind the objectives set by management.
4    Assisting with the implementation, documentation and maintenance of the new system.

As stated earlier, the precise nature of the systems analyst's activities varies between organizations. In a large company the analysts are normally organized into teams — each team concentrating on one or more applications. The more senior analysts act as team leaders so as to co-ordinate and guide the work of the less experienced members of the team. Where teams exist, it is essential that their individual efforts are co-ordinated at every stage, and not merely forced together after completion. This can be accomplished by the Systems Manager arranging regular team meetings, and by his encouraging continuous inter-communication between them.

### Business analysts/technical analysts

In large organizations it is sometimes advantageous to split the tasks of systems analysis into two parts. A business analyst is responsible essentially for defining the aims and problems of management and the organization. It is his duty to bring together the needs of the individual managers and departments in such a way as to enable an integrated and comprehensive system to be designed and implemented. This work calls for a high level of understanding of business in its entirety and, at the same time, a full appreciation of the problems entailed in transferring applications to computer-based systems.

In some instances a business analyst specialises in a certain application, e.g. reservation systems, and so works entirely within this work area. In other cases he works within the business as a whole; this obviously demands a comprehensive understanding of the organization's functions and objectives.

The former case means that the business analyst must have considerable experience of the application area, probably acquired through several years of specialising in the related problems.

The latter case also involves business experience and qualifications in applications such as accounting, industrial management, marketing and production control.

In either case a business analyst acts in effect as a consultant to top management and this implies that he needs to have gained deep and broad experience by having worked with several companies during his career. It is likely that a business analyst is professionally qualified in at least one aspect of business, e.g. cost and management accounting, production administration, or marketing.

A technical analyst works much closer to the computer in that he is concerned with translating the needs of management and the organization into a form that can be implemented on a computer-based system. If a business analyst is also employed then his problem definitions are converted by the technical analyst into computer-assimilatable forms. The technical analyst is therefore expected to know more of the technicalities of computer methods and rather less of the problems and aims of business. It is quite likely that the technical analyst has reached this position via several years as a programmer followed by a conversion course.

From the above remarks it is obvious that the work of the business analyst and the technical analyst overlap to a considerable extent. This is demonstrated by the fact that the majority of systems analysts fulfill both roles.

*Programmer/analysts*

Within a small company, and indeed in some larger organizations, the systems analyst may also perform the duties of programmer, in that he continues his systems design to the point of preparing the computer's detailed instructions. The efficiency of combining these occupations depends entirely on the qualities of the person involved. Provided he is of high enough calibre, both social and intellectual, to contend with the dual role, there are several advantages to be gained. The foremost of these is the elimination of the communication problem that can exist between the systems analyst and the programmer (Section 2.6). Another significant advantage is with regard to the flexibility of staff disposition; this is increased because each man can do either job. This versatility results in a smoother work load, and in a reduction of the idle time that can be caused through staff awaiting the output of work from each other. The main problem connected with the dual role is caused by the clash of the attributes desirable for each occupation. The systems analyst must be fairly extroverted and content to spend many hours talking to people in their 'language'. The programmer, on the other hand, must be prepared to spend even more hours working individually with the computer's language. It is unusual for both these sets of traits to be possessed by the one person.

## 2.3   Desirable knowledge and attributes of the systems analyst

A list of the desirable qualities of a systems analyst reads as a testimonial for both social and intellectual perfection. It engenders in its reader a scepticism towards belief in the existence of such a talented person. Nonetheless, these desirable qualities are worth reflecting upon and, if borne in mind, will help the prospective systems analyst to achieve his aims.

*Education*

This should be equivalent to at least a good GCE 'A' level, but higher academic qualifications than this are not absolutely necessary. An increasing proportion of systems analysts are graduates in one or other of a wide range of subjects and it is probable that in the future the majority of systems analysts will have degrees in either Business Studies or Computer Science.

## Background experience

Between the dates of qualifying educationally and starting work as a systems analyst it is very desirable that his background experience has been widened. Several years spent in a variety of companies or departments provide invaluable experience for this occupation. Ideally this background experience embraces work in financial and cost accounting, planning and administration. An extensive background of computer programming is of dubious advantage and, on its own, is not an acceptable qualification. In the past, results have shown that only about one in three good programmers have subsequently made good systems analysts. In a similar way a long period spent in one department only is likely to have instilled a blinkered outlook and to have made the person hidebound by his previous experience.

## Knowledge

The extent and pattern of the knowledge needed by the systems analyst depends largely upon his precise duties and upon the type of organization for which he is working. His knowledge need not necessarily include mathematics, but a comprehensive understanding of arithmetic and the ability to cope with figures are very desirable. Nor does it include formal qualifications in accountancy, but a good general knowledge of this subject is very useful.

Further desirable knowledge would include:

1    A broad understanding of business procedures.
2    The routines and techniques involved in all or some of the following:
Production planning and control.
Stores and stock control.
Accounting procedures.
Administration and O&M.
Marketing operations.
Preparation of surveys and analyses.
Operational research.
3    An understanding of the purpose and objectives of the organization from the viewpoint of its top management.
4    The techniques of systems analysis and data processing, including some familiarity with programming strategy and languages.

5    A knowledge of the range of data processing equipment (hardware) that is currently available and the software associated with it.

*Attributes*

All or some of the following attributes will prove to be useful:

1    A capacity to absorb quickly the wide concept of a system and also to give meticulous attention to its detail.
2    The ability to get along with all types of people and particularly with management, clerical and administrative staff, and programmers—to some degree the analyst must be 'All things to all men'.
3    The possession of a self-assured but not flamboyant manner.
4    A scientific scepticism, being interested in but critical of existing methods and information given to him. The ability to adopt a detached attitude is a valuable asset when inter-departmental discussions become heated.
5    Patience—a willingness to explain new methods, repeatedly if necessary, to other members of staff, including the programmers. Also patience in waiting for people to collect facts and for their ideas to mature.
6    A genuine desire to explore situations in order to discover the underlying factors, and an eagerness to learn new methods, techniques and developments in hardware.
7    A capacity to think logically combined with a 'fail-safe' mentality—this enhances good systems design, especially from the security aspect.
8    The ability to communicate verbally and in writing, and thereby transmit enthusiasm for new ideas and information about new systems.
9    Perception combined with tact, so as to be a good prober without, at the same time, giving the impression of grilling people.
10   A willingness to admit that certain problems reach beyond his knowledge and accordingly summon specialist assistance, for instance in the areas of operational research, statistics and financial modelling.
11   The ability to make good notes combined with a reasonably good memory. An exceptional memory is not necessary; the amount of facts involved would swamp the best of memories anyway.
12   Diplomacy—so as not to become regarded as the 'hatchet' man nor the 'efficiency' man.

## 2.4 Relationship between management and the systems analyst

The relationship between the systems analyst and top management depends largely upon the size and structure of the organization. In a large company it is obvious that every systems analyst cannot have unrestricted access to the managing director. In a small company, however, this situation may be both possible and desirable. Hard and fast rules cannot be laid down to cover all circumstances, but whatever prevails, a firm link must be forged between the planners and the decision makers. This link between systems analysts and management, if not direct, is via the data processing manager or the management services manager. Whosoever is involved as intermediary must be capable of appreciating the significance of the analyst's work so that he can support the resultant recommendations during top-level discussions. The link should be as short as is possible without invalidating the organizational structure of the company.

During the early stages of systems investigation a different situation may exist. At this point of time it is unlikely that management services and data processing will have been established. Their existence will probably be somewhat nebulous and consequently there is reason for more direct contact between the systems analyst and top management. In some ways the calibre of the systems analyst employed at the outset must be higher than that of later comers. Above all he must be capable of talking to top management in their own terms and of understanding their problems. His attitude should be such as to convince management that he is far more concerned with the advantages that data processing can bring to the organization than with computers in themselves. That is to say, he should be manifestly a company-man more than a computer fan.

## 2.5 Co-operation between the systems analyst and user departments

During the course of his investigations the systems analyst may find himself in the paradoxical position of requiring the utmost co-operation from some of the departments in order to bring about their decline. Whether or not this is the case, the advice of departmental staff has no substitute, and must be sought with diligence. The sum of their experience accumulated over the years provides the embryo of the new system. By realizing this, and accepting that he is not an expert on every subject, the analyst has taken a step on the road to successful systems design.

When discussing the work with departmental managers, the systems analyst's remarks should be directed towards making them feel that they will form an essential part of the new system. Before asking them for their additional information requirements and suggestions for improvements, it is advisable for the analyst to explain briefly the potentialities of a computer system. This puts the data processing picture in perspective and encourages the managers to discuss information requirements previously considered to be unattainable. It may take some little time for ideas to mature and so the systems analyst should allow time for deliberation.

Whilst giving every consideration to advice received, the systems analyst must guard against allowing outdated ideas to dominate his thinking. This philosophy applies particularly to systems created by the O&M department in the days before computers were taken seriously.

*Project teams*

A project team is a group of people working together in order to accomplish a predescribed task, i.e. a project. The project may form part of a systems investigation or the design or implementation of a new system. Thus, in a large organization several project teams work simultaneously on various projects. Each project team comprises from four to six members allied together so as to make the best use of their aggregate knowledge and experience. After the completion of the current project, the team may be dispersed, its members joining other teams or returning to their former roles.

It is likely that some of the staff from existing departments will be members of the project teams, especially during the systems investigation stage. These people obviously have a good understanding of present procedures, and can thereby make significant contributions to the aims of the new system. Generally it is necessary to give user-department staff some training in systems investigation methodology before they join a project team, and thereafter to guide their efforts for some length of time.

A project team may include business analysts, technical analysts, programmer/analysts and user-department staff, under the leadership of the most suitable person for that particular project.

The work of several project teams needs to be carefully planned and controlled so as to avoid duplicated effort and omission of necessary activities. Regular meetings should take place and all progress reported and monitored.

## 2.6 The systems analyst's role in the data processing department

In a large date processing department it is likely that the systems analysts will comprise one or more teams under the leadership of a senior analyst, who also acts as deputy data processing manager. Another arrangement is for the data processing manager to be in charge of programming and operating only, while the systems analysts form a separate department under the authority of a systems manager. This person reports directly to the management services manager or someone of similar status.

Whatever the arrangement, it is essential that the 'interface' between the systems analysts and the programmers is clearly established at the outset. Failure to do this can result in demarcation disputes arising during the programming and implementation phases.

What is meant by 'interface' in this context?

1   The precise level of information to be provided by the systems analyst, and the form in which it is to be presented to the programmer.

2   The authority of the systems analyst in relation to disputes about the precise methods and techniques to be adopted. Programming problems must not be allowed to dominate the methods employed in the new system.

3   The maintenance of contact between the systems analyst and the programmer during the programming phase. Contact should be close enough to ensure that the programmer keeps 'on target' and that no deviations creep into the program due to misunderstandings.

4   The degree to which programmers are permitted to contact staff outside the data processing department for additional information, and the extent to which they are allowed to make amendments to the routines as specified by the systems analyst. Within limits these functions are permissible for the more senior programmers, and there are two main advantages to this. First, in a situation where the systems analysts are hard pressed on further systems, they do not have to break off to deal with minor queries or changes. Second, it makes a natural introduction to systems work for programmers.

Although this interface should be established initially in a formal manner, it can be modified later when the parties have become more skilled and familiar with each other's capabilities. It is however advisable to preserve the formal definition of interface for the benefit of new data processing staff.


*Promotion prospects*

Within an established data processing department the systems analyst's line of promotion is normally via senior analyst to data processing manager or systems manager. Where the department has not yet been created, the systems analyst works alone and may become data processing manager designate if and when the need for data processing becomes apparent. Promotion to the position of computer operating manager would not suit the majority of systems analysts – this job is too routine for the explorative mind. In the larger organizations a line of further promotion is to management services manager, and from here his wide knowledge of the organization makes him eligible for the highest management positions. Experienced systems analysts will continue to be in demand during the foreseeable future, especially in the developing countries. They also have promotion prospects that are better than those of the majority of occupations.

## 2.7 References and further reading

2.1 'The job of a systems analyst', *Computer Bulletin* (December 1966).
2.2 'Education and training of systems analysts', *ibid* (June (1967).
2.3 'The dilemma of the systems analysts', *Computers and Automation* (August 1970).
2.4 'The role of the systems analysts', *Data Processing* (September/October 1968).
2.5 'Business education and the computer professional', *Computer Bulletin* (November 1970).
2.6 'Systems analysis – solving the manpower problem', *Data Systems* (December 1968).
2.7 *Systems analyst selection*, NCC (1970).
2.8 'Management training for d.p. people', *Data Processing* (March 1977).

# Chapter Three
# *The objectives of the system*

## 3.1 Immediate and long-term objectives

An analysis of the replies to the question, 'What are your immediate and long-term objectives?', would provide interesting if not altogether productive results. As might be expected, the replies would show a marked degree of correlation with the positions of the persons replying. Also they would contain a proportion of irrelevant information that may obscure the real objectives. The chairman of a company may have the immediate objective of satisfying the shareholders at the next meeting, and the long-term objective of taking-over a competitive firm. The production manager's objectives may be to reduce overtime working in the short-term and to re-equip the machine shop in the long-term. Whereas most members of staff will subscribe to the aim of profit-making or of meeting a pre-determined target of some sort, this is by no means their most exercised thought. Quite naturally their attitudes are coloured by personal ambitions, capabilities and anxieties, resulting in distorted objectives. The systems analyst, particularly if he is new to the organization, will not always find it easy to dissect and evaluate these diverse and conflicting aims.

In general, the higher a person's status the longer term are his objectives; and it is long-term objectives that form the true basis for systems planning. Although immediate objectives often tend to dominate the scene, these can be dangerous if they seriously conflict with the long-term objectives. An instance of this situation is where a computer has been installed at relatively high cost merely in order to counteract a shortage of skilled clerical staff. The short-term aim may have been achieved but the long-term objective of improved management information is frustrated by the clerically orientated computer system.

The system analyst's major problem is to balance the immediate and long-term objectives when designing a new system. All aims and requirements must therefore be weighted – qualitatively if not quantitatively – before using them as the basis for a data processing system.

## Financial objectives

Almost all organizations are in some way bound by financial considerations. These may be based upon the need to maintain profits, reduce costs or keep within a budget. Whatever the situation areas of high expenditure are of considerable interest in the early stages of systems development. If large sums of money are being spent, there is always a distinct possibility that costs can be reduced. Similarly if there is a large inflow of cash, there is likely to be a way of increasing it through the provision of better and more up-to-date information.

When considering financial objectives in relation to computer-based data processing systems, we must look forward over a period of at least five years. It is unlikely that a shorter period will give any time indication of the financial trend. Similarly, when comparing the potential savings and costs of several alternative systems, these should be discounted back to the present date (or some pre-decided future date). Without utilizing a recognized investment appraised method, we are not making a valid comparison, and therefore not truly assessing the likelihood of achieving the financial objectives.

Another point to bear in mind is that a data processing system that merely emulates a routine bookkeeping routine is unlikely to save large sums of money. Unless the cost of clerical labour is very high, the cost difference between manual and computer methods is unlikely to be other than marginal. It is the gains brought about by improved financial control through better information that will ultimately lead to reduced costs and increased profitability.

## Control objectives

Control of a company's or an organization's activities is achieved through the provision of suitable and timely information. This objective can be attained by means of computer-prepared reports distributed to the right people at the right time and containing all the relevant information. It is obviously important that the information comes directly and quickly into the hands of the person(s) who can take the appropriate control action, and that it contains sufficient facts to enable him to do so efficiently.

Control information falls into four main categories:

1    Reports of past activities or situations.
2    Analyses of past activities.
3    Predictions of future situations.
4    Plans for future activities.

*1    Reports*    A report gives details of what has actually ocurred and, as far as is possible, should contain information as to |why. The greatest dangers with regular reports of a routine nature |are that they either overwhelm their recipients with their enormity or are so uninteresting that their recipients take no heed of the contents.

*Exception reports*    These are prepared and/or distributed only when circumstances justify this. An exception report contains all the information appertaining to the exceptional activity or situation. It may also be beneficial to insert facts relating to other, non-exceptional, activities in order to provide a basis for comparison, the exception being highlighted in some way.

A difficulty with exception reporting is in ensuring that over a long period of time the exception parameters remain valid. This means either they have to be reviewed from time to time or they have to be flexible enough to cope with changing circumstances. Ideally, an exception parameter is self-adjusting based on other figures. A stock-holding report, for instance, highlights low-stock items. The level at which this is reported would usually be better decided in relation to the item's current demand than as a pre-decided fixed amount.

*On-demand (request) reports*    These are supplied only when requested by the recipient. It may be necessary to prepare on-demand reports in any case in order that they are readily available when requested. This is important as otherwise the information therein could be out-of-date by the time the report has been requested, prepared and delivered. If an on-demand report is not requested for a long time, it is possible that it is no longer of use and can be taken out of the system. With this possibility in mind, it is useful to make a note of the dates of the requests.

*2    Analyses*    An analysis is a summarized report of past activities. The data on which the analysis is based is aggregated into various totals so as to provide a more succinct picture of the whole situation.

One of the major problems in designing an analysis is to ensure that important factual details are not obscured by the aggregation. An exceptionally high item cost, for instance, could easily disappear into an aggregated cost total and thus not be detected. It is often advantageous to prepare an exception report along with an analysis.

Another aspect of analysis are the groupings of the aggregated figures. There are usually many possibilities in this respect, especially when the various combinations and arrangements of the groups are taken into consideration. Sales analysis, for example, are often based on what, when, where and to whom? This means that there are 64 fundamentally different analyses that can be prepared quite apart from their precise contents. For example: what/when/where, to whom/where/when/what, what/to whom/, and so on. If we also take into account the inclusions within these combinations, e.g. commodity groups or sales areas, the possibilities are almost infinite.

Ideally, analyses are designed so as to be flexible in their usage i.e. to provide information that is easily changed in its structure and content as and when necessary. The obvious advantages of such an arrangement have to be balanced against its cost in terms of computer programming effort. This introduces considerations as to the genuine need for flexibility as compared with a comprehensive but pre-determined type of analysis.

*3 Predictions* A prediction, often termed a forecast, is usually based upon a projection into the future of data pertaining to past activities. There are several sophisticated prediction techniques programmed for computer application, and consequently detailed predictions are easily and rapidly prepared.

Perhaps the biggest consideration is that the manager who makes use of a prediction fully understands what the figures in it actually mean and appreciates how they are derived. It is normally essential that the figures in a prediction are combined with other information before future planning and control decisions are taken. This information is usually based on either a knowledge of definite future events or subjective judgements founded on the manager's past experience of similar situations.

*4 Plans (schedules)* A manager's ability to plan future activities and the associated resources depends on the computing power at his disposal. In most circumstances there are many alternative arrangements for disposing his resources and scheduling the activities. Certain situations lend themselves to optimisation, i.e. to achieving the 'best' plan within certain imposed constraints. Examples are where operational research (OR) techniques such as linear programming, replacements, assignment, queueing and inventory control can be applied. In real life situations these OR techniques inevitably demand computing power as there are too many variables to be handled manually. OR is an extensive subject outside the scope of this book; the reader is directed to reference [3.3] for further information.

Planning also incorporates the more mundane problems such as factory loading, materials requirements and project control. The former applies to machines and labour requirements required to fulfill a workload and, in some cases, is amenable to solution by one or more of the aforementioned OR techniques. Materials requirements are based upon intended production programmes, and entail breaking these down into the constituent sub-assemblies, components and raw materials needed over the weeks of the programme.

Project control normally implies the utilization of network planning (PERT), and this again calls for considerable computing power. Network planning caters for the planning, monitoring and control of projects that comprise many interrelated activities, an example of which is described in Section 13.6

## 3.2    Top management support and involvement

For many firms the purchase of a computer is the biggest single financial transaction that its management has ever made. Although management might not be aware of it at the time, the decision to utilize a computer could be one of the most significant ever made within the organization.

In spite of these two important factors many managers have no appreciation of the computer's potential as a management tool. It has often been stated that lack of top management involvement is the most important contributory factor in the failure to use computers properly. In the early days of computers, management understandably adopted the attitude 'leave it to the experts'. At that time the experts were the programmers because they alone knew how to control the 'electronic brain'. What must nowadays be accepted is that it is the managers who should acquire the expertise, to obtain full value from data processing. The computer and the associated system are theirs, to be utilized to provide management information from the mass of routine data. This is not to suggest that management must learn programming; what is necessary is for them to understand the computer's potential and the principles of data processing systems.

With these aims in view, a worthwhile first step is for managers to attend a computer appreciation course. These are run by all the computer manufacturers and by a number of consultancy organizations and professional bodies. They usually involve full-time attendance at an education centre for a  few days. Although some of these courses may give the impression of attempting to indoctrinate, the experienced manager is able to see through this and perceive the real value of the subjects taught.

A second step is for management to give full support to an investigation of the existing system and to make a close study of its findings. This also involves them in directing the course of an investigation so as to steer it towards their own aspirations. The results of the feasibility survey (Section 3.3) provide an early indication of the course being followed, and by maintaining contact with the systems investigation and design, the course can be re-checked at regular intervals.

A third step for management is to make a clear decision regarding adoption or otherwise of the new system. This decision is based upon the systems analyst's report (Section 12.4).

A fourth step is to support, and to some extent become involved in, the implementation of the new system. By so doing they demonstrate that they have accepted the system and intend to make good use of it.

*Systems investigation support*

How can top management best give support at the investigation stage?

1    By providing the systems analyst with an assignment brief (Section 3.3), and being prepared to explain objectives, policy, and developments in and around the organization.

2    By officially informing all departmental managers of the impending feasibility survey and systems investigation, giving brief but convincing reasons of the purpose of these activities.

3    In helping to overcome any resistance to the investigation, and if necessary, using managerial authority to direct staff to co-operate. Also, where necessary, arranging for departmental managers to make available a few of their staff to assist temporarily in the systems investigation as members of project teams.

4    By appointing one member of the board (or its equivalent) to be responsible for management services, including systems analysis and data processing. The senior systems analyst and/ or data processing manager report to this director, who takes a particular interest in the top management aspects of data processing.

*The steering committee*

The board member appointed to be responsible for data processing

ideally should not have any bias towards a particular department.
The new system will cut across departmental boundaries and involve
integration of their work, at the same time requiring the utmost
co-operation from their staff. In order to meet these requirements
it is advisable for a steering committee to be formed with the above-
mentioned director as its chairman.

Other members of the committee might well be the following:

Management Services Manager (when appointed).

Data Processing Manager (when appointed).

Senior Systems Analyst (until appointment of either of the
above).

Managers of the departments most affected.

Representative from outside organization such as computer
manufacturer or consultancy firm (if involved).

At certain times other persons can be co-opted on to the committee
when the areas of work with which they are concerned are under
review. As with all committees it is advisable to restrict the membership to not more than about eight persons so as to facilitate
decision-making. The decisions made by the steering committee
are those that affect the company from the data processing aspect.
Decisions appertaining only to the internal working of the data
processing department can be made without reference to the steering committee.

The terms of reference of steering committees vary slightly from
one organization to another, but the main points are outlined
below:

1    To establish a programme of activities connected with the
     actual or possible introduction of data processing into the
     organization. This programme includes the systems investigat-
     ion, a study of its findings and the recommendations there-
     from, the systems design, and the implementation of the data
     processing system.

2    To maintain a watching brief on the above activities, and to
     use its influence in overcoming problems of an inter-depart-
     mental nature.

3    To advise on the possible purchase of a computer, and to
     select firms to be invited to tender, thereafter acting as the
     official negotiating body with prospective suppliers of all
     data processing equipment.

4    To appoint a data processing manager if one is not already
     designated, and to provide his terms of reference.

3.3    The assignment brief and feasibility survey

An assignment brief and the consequent feasibility survey are inevitably tied together in that the one always has some effect upon the other. It is unrealistic to assume that a precise assignment brief can be drawn up to cover all the possible points that can arise. It may be created prior to the formation of the steering committee so that it is the responsibility of top management to prepare it. The problem here is the unlikelihood of their appreciating the potential of data processing until the feasibility survey has been carried out. Thus there must be an iteration between the assignment brief and the feasibility survey, the former being amended perhaps several times as a result of the findings of the latter.

*The assignment brief*

The major features of an assignment brief are as follows:

1 It is an authorization for the systems analyst to carry out an investigation of existing systems. In this respect it is a 'passport' into all relevant departments, and should be regarded as a request from top management asking staff to assist the analyst in his investigations.

2 Initially it states in general terms the objectives of the feasibility survey, then, after suitable amendments have been made, it leads to the detailed objectives of the full systems investigation.

3 It indicates any limitations to be imposed on the survey – these may refer to applications, areas of work, or locations. Limitations should not be imposed without good reason because they can severly restrict the potential benefits of a system. Any areas that are not specifically prohibited are assumed to be within the scope of the survey.

4 Reference is made in the assignment brief to any previous surveys that have been made in a similar field. It is unwise for top management to deliberately conceal the results of previous surveys even though they are outdated and were unacceptable at the time. Impractical suggestions contained therein might now be entirely feasible and beneficial.

5 It provides an indication of the top limit cf capital expenditure or annual outflow that can be considered. This limit must be realistic in relation to the hoped-for savings in future years, and is one of the features most likely to be amended as a result of the feasibility survey.

*The feasibility survey*

This forms an interim stage between the assignment brief and the full systems investigation. In certain circumstances it is permissible to omit the feasibility survey and to proceed immediately with the full investigation. This is usually the policy when the area of the investigation is limited or when it is the intention to transfer a well-understood procedure on to an established data processing system.

Generally the feasibility survey is carried out by the systems analyst who will later be engaged in the full systems investigation and design. The time taken for the survey should be kept quite short subject to achieving the required results. During the survey the main source of information is line management.

The main aims of the feasibility survey are as follows :

1  To determine whether the objectives stated in the assignment brief are attainable, and if not, what constraints must be removed.

2  To define the major problems existing within the organization so that the systems analyst can plan his strategy for the full investigation. In a large organization this includes the setting up of project teams consisting of systems analysts and staff recruited temporarily from the prospective user departments. The feasibility survey provides a good opportunity for selecting suitable persons for this work.

3  To find the areas where potential exists for making savings—of money, time or effort. These areas may not necessarily be considered to be problem areas but are usually those in which high expenditure is incurred.

4  To decide if specialists will be needed to render assistance in the full systems investigation. These people may be enlisted either from within the organization or from outside agencies such as consultancy firms or software houses. The problems requiring specialized knowledge include those involving operational research techniques, statistical methods, mathematics and investment appraisal techniques. Apart from those needing specialists, the systems analyst may come across simpler problems that merely call for a revision of his knowledge or a small amount of additional training.

5  To approximate the time required and the cost of the full investig-ation. The time required is dependent on the number of systems analysts who will be available, and if additional analysts are needed, their numbers and calibre should be decided at this stage.

In view of the restricted time allowed for the feasibility survey, grandiose reports are best avoided. It is satisfactory to report the findings in the form of an extended memorandum combined with a verbal explanation of any ambiguous points. A detailed report is, of course, presented to top management after the full investigation has been completed (Section 12.4).

The memorandum should cover the aims of the survey together with a conclusion in which are recommendations for further action. Upon rare occasions nothing is to be gained from further investigation since it is perfectly clear that the existing system cannot be improved upon. An appropriate recommendation in this situation is that another feasibility survey is carried out either at a stated date or if certain circumstances arise.

### 3.4   References and further reading

3.1    'Computers and the confused MD', *Data Systems* (June 1973).
3.2    'Management implications of DP systems', *Computer Bulletin* (February 1971).
3.3    MOORE, *Basic operational research*, Pitman (1976).
3.4    'What management expects of EDP and vice versa', *Business Automation* (1 February 1971).
3.5    'Some problems of user management', *Data Processing* (September 1970).
3.6    'Lies, damned lies and feasibility studies', *Data Systems* (March 1970).
3.7    'Mount that tiger', *Data Processing* (March 1971).
3.8    HAROLD, *Orienting the manager to the computer*, Auerbach (1972).
3.9    'A survey of management information systems literature', *Computer Bulletin* (June 1971).
3.10   'Solving the information maze', *Data Systems* (May 1970).
3.11   'Where are the power centres?', *ibid.* (February 1972).
3.12   'Computers and the ecology of management', *Data Processing* (July/August 1972).
3.13   'Some considerations of the cost and value of information', *Computer Journal* (May 1969).
3.14   'Computer in mainstream management', *Data Systems* (July 1968).
3.15   *Using computers: a guide for the manager*, NCC(1971).
3.16   'Do you know what you want?', *Data Processing* (May-June 1971).

# Chapter Four
# *Systems investigation*

When the outcome of the feasibility survey is known, top management is in a position to decide whether to authorize a full investigation of existing systems. If this is the decision, approval should be made official by publishing a memorandum or notice to this effect, after which the investigation can proceed.

At this stage the systems analyst is probably confronted by a large organization whose complexities can appear overwhelming. A full investigation can, if completely unrestricted, take an interminable amount of time and effort. The systems analyst, by using the information obtained during the feasibility survey combined with his own experience, will impose his own limitations on the depth of the investigation. This is necessary in order to obtain the required information within a reasonable period of time, and without accumulating a large amount of irrelevant data. Although the investigation must not be too protracted, an attempt to work to a tight time schedule would almost certainly fail. This is because the systems analyst is in the hands of other people; it is their availability that largely determines the rate of progress.

Where the organization under investigation is large, the systems investigation must obviously be carried out by several systems analysts and, possibly, by teams of analysts. In any event, the work of investigating existing systems is greatly facilitated by the employment of staff experienced in the business of the company. When the systems analysts themselves do not have this facility, it is advantageous to engage experienced user department staff temporarily as members of project teams (Section 2.5). This arrangement means that close contact is maintained between the systems and user departments, and engenders a feeling of involvement amongst the prospective users, thus encouraging their acceptance of the new system.

## 4.1  Fact finding and verification

*What are facts?*

The definition of 'fact' in the scientific sense brings in the meaning of truth and accuracy. For the purposes of data processing, however, facts can be regarded as any information that is relevant to the existing or proposed system. Our main concern is to be able to find all the facts, to separate them from opinions, and at the same time not to lose sight of interesting and potentially valuable ideas. Facts are the 'bricks' from which the new system will be built, ideas are its architecture. Provided they are not confused in the mind of the systems analyst, both are of value in the design of the data processing system.

The systems analyst is also concerned with the accuracy of the facts given to him. In this context accuracy can be thought of in two ways – is a statement true or false – how near the absolute truth is a figure of measurement? These are, of course, metaphysical ideas and the analyst is more concerned with the mundane task of gathering 'approximate' facts than in philosophising about their absolute accuracy. Whenever a figure, such as the number of employees in a department, is ascertained, it is subject to a degree of inaccuracy. Employees come and go, so there is no point in trying to obtain absolutely accurate information of this nature. All that is of interest is that there are about 23 employees, for instance, in the department and, provided this figure is not likely to change considerably, it is satisfactory for the systems analyst's purposes.

There are, on the other hand, certain facts whose accuracy must be absolute, and others for which a maximum allowance can be made. The former relate to things that control the design of the data processing system; for instance the layout of a code number because this may be used to decide the technique to be adopted for accessing stored data. The 'maximum allowance' facts are those such as quantities, prices and rates, that are to be processed by the computer. Because we are dealing with a dynamic organization and transient situation, these facts are nowhere near constant; the best we can do is to make allowance for the maximum value that they can attain.

*How are the facts gathered?*

The facts and information required from the investigation are not always in a written nor immediately usable form. It is sometimes necessary to create or deduce the required information from other ascertainable facts. Generally this is a simple task involving nothing more than the addition of a few figures. To the existing system these deduced facts may be of no value and therefore are not readily available. For example, the total number of bought parts handled in a factory has not been previously ascertained because the existing production control system deals with them in quite separate groups.

Facts are obtained by four main methods – asking, observing, measuring and reading.

At the outset of the investigation the systems analyst must prepare the ground so that he is able to complement these methods in an efficient and logical way. In order to gain access to the appropriate people and places, he should start by approaching the departmental managers. Thereafter, as far as is possible, all contacts should be made via a more senior person; if necessary going back to top management for help. In most cases this will not be necessary because sufficient contacts will have been established during the feasibility study.

Having gained access to a department, the systems analyst should break down the psychological barriers that are bound to exist between himself and the department's staff. The surest way of doing this is for him to become acquainted with all the staff. With this in mind, the analyst is well advised to obtain a temporary working place near to the department's centre of activity. This not only creates a rapport between himself and the staff, but also facilitates observation of the department in action.

*Asking*　This usually entails the verbal questioning of the appropriate members of staff, and carefully noting their replies. A reply might not be strictly pertinent, in which case a re-phrasing of the question is called for. It is imperative that the systems analyst has sufficient knowledge of the topic under discussion to detect a divergence of understanding.

As well as asking about the actual details of the present system the analyst ought to obtain ideas and suggestions for improvements. These are then mentally sifted in order to separate mere wishful thinking from the seeds of genuine advancements. A little gentle probing often encourages people with valuable ideas to make them known. The feeling that changes are in the offing, and the presence of an attentive listener, bring forth suggestions that might otherwise never be heard. Persons who seem to be completely devoid of ideas, if given the stimulus and some time to think, can sometimes produce previously suppressed but valuable ideas.

*Questionnaires*　An alternative means of asking is the questionnaire. This method is suitable for situations in which a large number of people are to be asked a number of straightforward questions. These would call for brief answers that can be recorded by the recipient of the questionnaire in a form that precludes subsequent misinterpretation. The questionnaire must be carefully designed so that the nature of the required information is clearly indicated. Great care is necessary in the phrasing of the questions, and they should be accompanied by a brief explanation of their general purpose.

This method is unsuitable for eliciting facts related to an actual system; the questions are too limited, and the replies often not forthcoming for a considerable time. A possible application of the questionnaire within systems investigation is as a means of interrogating widely deployed staff. These people might, for example, be employed in the company's branches or depots in various parts of the country or the world. In this case a questionnaire can be used if the questions are suitable, and the required information is not available from a single source.

The following points should be borne in mind when designing a questionnaire:

1   The levels of intellect and interest of the recipients.
2   The incorporation of straightforward multiple-choice answers.
3   The desirability of identifying the respondent, especially if a complete set of replies is required.
4   Encourage the recipients to respond by explaining the survey's purpose and by facilitating their participation, e.g. by enclosing a prepaid addressed envelope where necessary.
5   Consider the means of analysing the replies if these are extensive in number. This could be done by using OMR questionnaires (Section 10.5), and analysing the answers by means of a computer.

*Observing*   It is an interesting scientific theory that the observer of a system influences its behaviour. This can be as true for social systems as for atomic. The less of an observer, and the more of a participant, that a systems analyst can become, the more realistic will be the picture he obtains. His acceptance as a temporary member of a department will enable him to observe while being part of the system. This facilitates a study of the actual situation, the department in action, and the methods for dealing with exceptional conditions.

Actual observation also gives the systems analyst the opportunity to assess the following characteristics of a department.

1   Pressure of work—high or low, steady or variable, isolated or evenly spread between the staff.
2   Movement of personnel—within the department and between it and other departments. This movement may be a means of transmitting information between departments (Section 4.3).

3    Attitude of the staff towards the existing system — favourable of dislike, of resigned acceptance, or possibly of not really understanding it.
4    Volumes of telephone calls, callers from other departments, visitors from outside the company and other interruptions to routine.
5    Usage of files—for routine purpose, answering queries, or other special reasons.

*Measuring*    The main purpose of this activity is to approximate the numbers of|documents, items, persons, transactions, etc., concerned with each sphere of work. These amounts are normally obtainable from the appropriate staff, but when they are unobtainable in this way, or are suspect, actual measurements or estimations must be made. If this is done by sampling, care should be taken to ensure that truly representative samples are used (see [4.12]).

Another purpose of measuring is to obtain the times taken for certain activities and their frequency of occurrence. These times include not only the actual task performance times but also the intervals between the completion of a task and its effect elsewhere. It is not recommended that all tasks and intervals are carefully /timed; the need is for the approximate times related to the main activities.

The frequency of activities applies in two respects. First, the frequency with which a routine is carried out—for example, invoicing is daily, payroll is weekly. Second, the rate of occurrence of variable events—for example, the average and peak rates of arrival of customers' orders.

*Reading*    Much of this topic is covered in Section 4.2 (Inspection of Documents), but reading also includes other information mentioned below.
1    Reports of previous surveys and investigations. These are worth studying carefully but any facts therein that appear to be usable should be confirmed by other methods. It is not satisfactory merely to inquire about the general veracity of a report; its originators are almost certain to confirm this regardless of changed circumstances. Recent reports are the most helpful but outdated ones may also provide useful ideas.
2    Company instructions, memoranda, and letters. These are likely to be voluminous; it is therefore wise to persuade others to extract the appropriate documents from their files.
3    Company information booklets and sales literature. These are worth scanning in order to obtain general information about the company's activities and connections.

## Verification of facts

Ideally the systems analyst obtains verification of every fact that he collects. In practice this is not always possible but nevertheless he must attempt to do so for all facts obtained verbally. The answer to a question can be erroneous for one of three reasons:

1    Ignorance of the subject leading to a mistaken answer.
2    A misunderstanding of the question provoking an irrelevant answer (sometimes not discernable as such).
3    A deliberate misrepresentation of the situation.

The third of these reasons is fairly uncommon and is not usually associated with quantitative facts.

How can these difficulties be dealt with? The obvious method is to ask another person and then compare the two answers. If the answers seriously conflict, their reconciliation requires a little diplomacy on the part of the systems analyst in order to avoid exhibiting someone's ignorance or mistake. Directly quoting the other person's answer is not altogether wise; the question should be posed in a non-leading form.

Another method of verification is to ask a more detailed question following the answer to the previous question. The answer may provide valuable information in itself but, in any case, should help to demonstrate the accuracy of the previous answer.

When a large amount of important information has been obtained verbally, written confirmation should be requested. By doing this informally and explaining the reason, the recipient of the request is not made to feel in any way disturbed.

In addition to the verbal means of verification, the systems analyst can sometimes employ one or more of the other three methods mentioned previously, i.e. observing, measuring, reading. As a general principle two of the four methods should be combined in order to obtain and check any set of facts.


## What facts are needed?

If the range of organizations and applications amenable to data processing were more restricted, a compendium of suitable questions could be usefully compiled and utilized in systems investigations. This approach is unsuitable for most situations but a check list might be helpful to a novice systems analyst.

The required facts, in addition to those that are described in Sections 4.2 and 4.3, can be broadly categorized as under.

*1 The sets of entities* pertaining to the organisation. In this context, entities include goods sold, parts manufactured, persons employed, suppliers, customers and so on.

> How many different entities in each set?
> How and why are these sets sub-divided?
> Is the set subject to rapid change in its size and content?

## 2 Information about entities

> What descriptions, quantities and values are associated with
> each entity?
> Are the descriptions standardized?
> What are the maxima and minima of the qualities and values?
> Which of the quantities and values are liable to fluctuate and
> what causes this?

In the above context 'descriptions' include names and addresses as
well as the conventional meanings. 'Quantities' covers stock levels,
order quantities, sales tax (VAT) percentages, discount percentages,
re-order levels and any other non-currency figures. 'Values' include
all amounts expressed in currency such as prices, wages, income tax,
costs, etc.

## 3 Code numbers in use
Normally each entity in a set has a code
number, although not all sets are coded. The information of interest
about each set is the precise layout and meaning (if any) of the code
numbers therein. A meaning is sometimes ascribable to a particular
digit or character in a code number, e.g. part numbers starting with
a five apply to bought-out components.

> To what extent (if any) the code is used outside the
> organization, and is there scope for this to increase? For
> instance the use of commodity codes by customers on their
> orders.
> What would be the effect of re-designing the set of code
> numbers? (This is normally a hypothetical question but it
> is just possible that this activity may have to be carried out.)

## 4 Calculations performed
Business calculations are not usually
complex but can be regarded as including all operations involving
quantities and values.

> What fractions and decimals are in use in connection with
> weights, lengths, areas, volumes, quantities and money?
> Do these pose any special problems connected with various
> currencies and units of measure?
> When extending (multiplying), how is rounding-off performed,
> and does this involve any reconciliation problems?

## 5 Amounts of data handled

> What amounts of data are received from and dispatched to
> outside organizations and internal departments?

Are these amounts reasonably constant or do they fluctuate significantly, if so, to what extent and in what way? Fluctuations may be seasonal, monthly, weekly or any other regular cycle, or at random.

Are the amounts of data increasing, if so, at what rate and is this likely to continue?

Where it is the intention to operate the new system on a real-time basis, the timings associated with the data play a more prominent role. Additional questions have to be asked and the replies verified carefully—actually measuring the relevant data if at all possible (for further details see Chapter 39 of reference [4.1]).

What are the rates and arrival patterns of each type of data, i.e. transactions or messages?

When do peak periods occur, for what reasons, and what levels of message rates are then reached?

How long is each type of input message and what amount of output information or length of reply message will it cause to be generated?

How quickly does the user require a reply to his messages, i.e. what response times are needed?

Real-time systems incorporating man-computer dialogues (conversational mode) need more information regarding the exchange of messages between the terminal user and the computer. The systems analyst needs to find what are the structures of the separate messages that will occur during a conversation, and the frequency with which these messages will be transmitted each way. It is unlikely that this information will be forthcoming merely as a result of asking for it. The prospective user of conversational mode needs to have the general procedure carefully explained to him and then to devote considerable time, in conjunction with the systems analyst, thinking about the precise dialogue he wishes to use (see reference [4.13] for further information on man-computer dialogues).

*Irrelevant data and information*

Desire on the part of the systems analyst to make the investigation thorough and comprehensive, engenders the collection of large amounts of data. Some of this is bound to be irrelevant, and an awareness of this **possibility enables the analyst** to avoid irrelevancies, thereby not only saving time and effort but also avoiding the obscurement of significant facts upon which the new system will be based. During the investigation the analyst is interested in information about the data rather than in the data itself. A study of a firm's products, for instance, should yield quantitative and qualitative information of their classification, manufacturing control, sales turnover and so on, rather than merely a list of the products.

Information regarding methods and policies is not easily segregated into 'wheat' and 'chaff' — today's wheat becomes tomorrow's chaff and vice versa. A piece of information that is insignificant in one situation can be of vital interest in another. For example, the impending retirement of the cost accountant may well become a valuable piece of information when considered in relation to plans for a new costing system.

With the gaining of experience, the analyst can avoid the accumulation of trivia without, at the same time, losing any essential information.

## 4.2 Inspection of documents

The term 'documents' as used in this section refers to anything on which facts and figures have been written or printed. It includes ledgers, lists, files, catalogues, interpreted punched cards, forms, and other documents of a similar nature.

*Meaning of entries on documents*

During the course of an investigation the systems analyst acquires copies of a wide variety of documents of various sizes, shapes and colours, and which contain a multitude of different entries. It is therefore imperative that all these entries are discerned and understood not only at the time of acquisition but also later when the new system is being designed. The majority of documents start life as blank sheets containing only their pre-printed headings. With the passage of time the changes inherent to any organization cause some of the headings and their associated entries to be at variance. To a comparative stranger to the system these variances may not be at all obvious; the systems analyst must therefore be watchful for :
 Entries under incorrect headings
 Entries never made despite the headings.
 Entries made for which no heading exists.
Detection of these irregularities is facilitated by inspecting 'live' documents holding real data. These can be inspected at leisure if photo-copies are made or if speciments of recently used documents are obtained. By inspecting live documents the systems analyst can judge their legibility, bearing in mind their possible future use as source documents. He should however carefully check the vintage of supposedly live documents.

## Document specification form

When inspecting documents the systems analyst should find the meaning of any entry that is not obvious and make a note of any special symbols, entries in coloured writing and remarks. For all but the simplest documents it is advisable for him to fill in a 'Document Specification Form' as a means of accurately describing the document's entries. An example of this form is shown in Figure 4.2; this has been filled in as for the order sheet in Figure 4.1. Most of the entries on the document specification form are self-explanatory but a few need a little more explanation.

*Reference No.*   This is any reference appearing on the document that identifies it; if there is none printed, an arbitrary number should be entered by the systems analyst on his specimen so as to connect it with its record specification form.

*Entry Ref.*   This is a reference written and circled on the specimen document by the systems analyst so as to identify each entry thereon.

*Picture*   The figures and letters in this column indicate the size and layout of the entry, each character has the following meaning:

| | |
|---|---|
| 9 | means a numeric digit |
| ½ | means 0 or ½ |
| ¾ | means 0, ¼, ½, or ¾ |
| A | means an alphabetic letter from A to Z |
| B | means blank |
| . | means decimal point |
| X | means any of the above and also symbols |
| 9(5) | means 99999 |
| X(4) | means XXXX, XXX, XX, or X |

*Examples of pictures*

| | |
|---|---|
| X(15) | means a field of up to 15 alphabetic letters, numeric characters blanks or symbols, e.g. |
| | JOHN HARGREAVES |
| | or    ACACIA AVENUE |
| | or    SHOE/BROWN 8½ |
| 999.9 | means a field of three whole numbers and one decimal, e.g. 125.7. |
| A(6) | means a field of up to 6 alphabetic letters, e.g. WASHER or BOLT. |
| AAAB99A | means three alphabetic letters, a space, two numeric digits and finally one alphabetic, e.g. FWD 36C. |

Occasionally several pictures may be applicable to one type of entry, e.g. the various layouts used for car registration numbers; in this case they should all be shown, with a brief explanation (if any) of each picture under 'remarks'.

*Maximum*   This is simply the highest value that an entry can attain.

**ORDER FROM:**
Pelham Caravans Ltd.
Pelham Street,
Newark, Notts (A)

**INVOICE TO:**
Same

**DELIVERY ADDRESS**
Pelham Caravans Ltd.
Mills Drive,
Farnden Road,
Newark, Notts. (C)

**CUST. ORDER NO.** Verbal (D)    No.3590 (G)

**TELEPHONE NO.** 0602-64387 (E)    SALESMAN'S SIGNATURE JW (H)

**CONTACT** Mr. Pollard (F)    DATE 28/3/78 (I)

**ACCOUNT** Monthly (J)    **NO. OF INVOICES** 1 (K)    **DELIVERY METHOD** O/V

Aluminium Alloy HE 30WP (M)

| Item (P) | Quantity (Q) | Description | Our Price (R) | Weight (S) | Supplier (T) | Our Order No. (U)(L) | Supplier's Price (V) | Supplier's Delivery (X) | Our Invoice No. (Y) | Invoice Value (Z) |
|---|---|---|---|---|---|---|---|---|---|---|
| | 2 | 6.9m SR12 Side Rave (Flat) (N)→(29) | 48½p | 2×6.9m 30kg | Impalco | 682 | 47p −5% | x/s | | 14.55 |
| | 2 | 7.2m 10×5cm lipped channels (41) | 47p | 2×7.2m 40kg | Impalco | (W) | 45½p −5% | x/s | | 18.80 |
| | 2 | 22×1cm corrigated planks (24) | 54p | 25kg 3×4·5m 41kg | Impalco | | 52p −5% | x/s | | 13.50 |
| | 3 | 4.5m 7·5×4×0·5cm channel (164) | 43p | 3×4·8m 44kg | Impalco | | 40p −5% | x/s | | 17.63 |
| | 3 | 4.8m 7·5×4×0·5cm channel | | 3×4·8m | Impalco | | | x/s | 1359 | 18.92 |
| | 4 | 4·8m 2·5×3×0·3cm angle (20) | 52p | 10kg | Impalco | | 49p −5% | x/s | | 5.20 |
| | | | | | | | | | (AA)→ | 88.60 |

Figure 4.1  Order sheet (specimen document)

In many cases it turns out in practice to be the same as the picture, but is sometimes less than this, e.g. entry reference K (number of invoices) has a picture of '9' but a maximum value of only five; 'maximum' is not really meaningful for non-numeric fields.

## 4.3 Information flow

The transmission of information within an organization is the means by which control and stability are achieved. It is therefore vital that no existing channels of communication are overlooked during the investigation of the present system. Since the main method of transmitting information and data is by means of documents, a careful study of document movement is essential.

*Document movement*

Documents originate, both within and outside the organization, at the time when the first entries are made on them. Thereafter further data is recorded and extracted at several points before they are finally filed away, destroyed, or dispatched to outside organizations. By tracing these movements and recordings, the systems analyst can build up a picture of the information flow within the organization. During his visits to the various departments, the following questions must be answered.

What documents originate here and how may copies are prepared?
To where are these documents dispatched?
What documents are received from elsewhere?
What documents are filed and for how long are they filed before being disposed of?
What entries are made on each document and what work does this involve?
What information is extracted from each document and for what purpose?

By cross-checking the answers received from the various departments omissions and misunderstandings are minimized.

*Document movement form*   This form is complementary to the document specification form (Section 4.2) but is less suitable for standardization. It is best designed to meet the precise needs of each company's document movement pattern, and is likely to contain all or some of the following information:

1   *Identification*   The document's name, title and/or reference number (tying in with the corresponding document specification form).

# DOCUMENT SPECIFICATION FORM

| NAME OF DOCUMENT | Order sheet | DEPARTMENT | Sales |
|---|---|---|---|

| REFERENCE NO. PFC 9171 | DATE FILLED IN 16.5.78 | FILLED IN BY HDC | |
|---|---|---|---|

| ENTRY REF. | HEADING | PICTURE | MAXIMUM | ENTERED BY | REMARKS |
|---|---|---|---|---|---|
| A | Order from | X(30) per line | X(100) in all | Salesman | Up to 5 lines |
| B | Invoice to | X(30) per line | X(100) in all | Salesman | Usually same as 'A' |
| C | Delivery address | X(30) per line | X(100) in all | Salesman | Usually different from 'A' |
| D | Cust. order No. | X(10) | X(10) | Salesman | Usually blank (verbal) |
| E | Telephone No. | 9(5) | — | Salesman | Picture varies somewhat |
| F | Contact | A(20) | A(20) | Salesman | Name of a person |
| G | No. | 99999 | 99999 | Pre-printed | Sequential ref. No. |
| H | Salesman's sign'r. | | | Salesman | |
| I | Date | 39.19.99 | 31.12.99 | Salesman | Date order received |
| J | Account | A(7) | A(7) | Salesman | Monthly or cash |
| K | No. of invoices | 9 | 5 | Salesman | = No. of order sheets |
| L | Delivery method | X(4) | See remarks | Salesman | Limited range (O/V, BR, BRS, Coll) |
| M | Description | X(30) | X(30) | Salesman | Isolated description refers to items following |
| N | Length (per piece) | 99 | 15m | Salesman | Circled figure = est. weight |
| P | Item | B | | | Not used |
| Q | Quantity | 999 | 500 | Salesman | Pieces ordered |
| R | Our price | 99½p | 75p | Salesman | Per kg. |
| S | Weight (actual) | 999 | 200 | Cost clerk | From which inv. val. calc. |
| T | Supplier | A(10) | | Purch'g clerk | Limited (could be coded) |
| U | Our order No. | 9999 | 9999 | Purch'g clerk | |
| V | Supplier's price | 99½p | 70p | Cost clerk | Per kg. ± discount |
| W | (Discount) | ±9½ | ±7½ | Cost clerk | + if surcharge |
| X | Supplier delivery | X/S or 99 | 15 | Purch'g clerk | X/S = immediate, or weeks |
| Y | Our invoice No. | 9999 | 9999 | Cost clerk | |
| Z | Invoice value | 999.99 | £400 | Cost clerk | = item value |
| AA | (Total value) | 9999.99 | £2000 | Cost clerk | On last sheet only |

**Figure 4.2**  *Document specification form*

2    *Purpose*   Concise details of its main uses.

3    *Origin*   Where and by whom it is originated, the number of copies made, and the differences (if any) between them.

4    *Distribution*   To whom each copy of the document is distributed after origination, and brief details of its movement thereafter, referring to the relevant flow charts (see below).

5    *Volumes*   The average and peak rates at which the document is created, and the average number of entries on it if these are variable. The number of documents in existence in each department at any point of time. A note of any cyclic or seasonal variations in the above figures.

6    *Sequences*   Any significant sequences in which the documents are moved or are filed in each department; these may arise naturally or through sorting.

7    *Special considerations*   Any further points of importance not included above.

## Flowcharts of existing routines

During the investigation the systems analyst pieces together the pattern of operations within each department's routines, and relates this to the movement of documents between departments. This pattern can be represented in a number of ways, but for most cases it is best done in diagrammatic form. This entails preparing a 'flowchart' such as that shown in Figure 4.3. This example relates to the order handling routine of a firm of aluminium stockists, and ties in with the 'order sheet' described in Section 4.2 and Figures 4.1 and 4.2.

On a flowchart each operation within an existing routine is represented by a box which is numbered arbitrarily for reference purposes. Within each box is a brief explanation of the operation together with the names and/or reference numbers of any documents involved. If the operation is too complex to describe adequately within the confines of a box, this can be done on a separate sheet and referred to in the box. Arrows indicate the general sequence of the operations, and allow for cross-referencing between flowcharts. The precise layout of the flowchart is not as important as its clarity, and provided all the systems analysts in the one organization follow the same conventions for drawing and numbering, no confusion will arise. Each flowchart has a reference number and it is often convenient to code this so as to indicate the department (or application) to which it applies. The flowchart in Figure 4.3 for instance has the reference number S4, the S indicating that it applies to a sales department routine.

A number of different conventions exist for flowcharting existing routines; these mostly differ only in the symbols used for the various

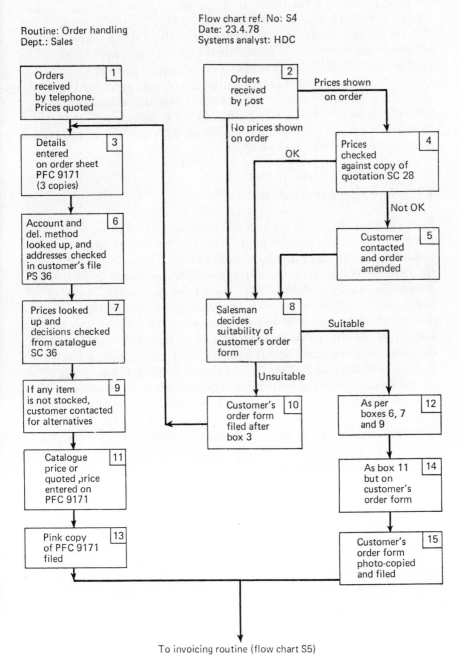

Routine: Order handling
Dept.: Sales

Flow chart ref. No: S4
Date: 23.4.78
Systems analyst: HDC

**1** — Orders received by telephone. Prices quoted

**2** — Orders received by post

Prices shown on order

No prices shown on order

**3** — Details entered on order sheet PFC 9171 (3 copies)

**4** — Prices checked against copy of quotation SC 28

OK

Not OK

**6** — Account and del. method looked up, and addresses checked in customer's file PS 36

**5** — Customer contacted and order amended

**7** — Prices looked up and decisions checked from catalogue SC 36

**8** — Salesman decides suitability of customer's order form

Suitable

Unsuitable

**9** — If any item is not stocked, customer contacted for alternatives

**10** — Customer's order form filed after box 3

**12** — As per boxes 6, 7 and 9

**11** — Catalogue price or quoted price entered on PFC 9171

**14** — As box 11 but on customer's order form

**13** — Pink copy of PFC 9171 filed

**15** — Customer's order form photo-copied and filed

To invoicing routine (flow chart S5)

**Figure 4.3** *Flowchart of existing routine*

43

types of operation (see [4.14]). Since no convention has been adopted universally the symbols of the convention adopted should be explained in a legend distributed to all users of the flowcharts. In any event, when drawing a flowchart it is advisable to allow plenty of space and not put over many operations on one sheet. This enables insertions and amendments to be made later.

The flowcharts prepared during the investigation of existing systems should not be confused with the systems design flowcharts drawn later (Section 10.7). The latter, although the same in principle, are drawn using standard symbolic shapes to represent the various types of computer and manual operations.

## 4.4 Detection of exceptions

Even the most experienced systems analyst is never quite free from the nagging anxiety that he might have overlooked an exceptional condition, or an exception to the exception. The people who explain the existing system to him naturally tend to concentrate on the normal and more straightforward aspects of their work. A verbal explanation, if broken up by discussions of special conditions, quickly loses its continuity and meaning for the listener. It is therefore advisable for the systems analyst to re-state the explanation, at the same time inquiring about exceptions at appropriate points. It may be necessary to do this several times, steadily filling in the details to complete the picture. It is not satisfactory to ask merely 'Are there any exceptions to this?'; but to select particular points of doubt and to pose questions in the form, 'What happens if . . . ?' This approach brings unusual conditions and circumstances to the mind of the interviewee; alternatively it may trigger off a new line of thought about the system.

In addition to verbal discussion, time spent in browsing through past documents often highlights former exceptional conditions that have been forgotten. These, however, may have been temporary and will not re-appear; but before disregarding them this must be established beyond doubt.

A complete list of all categories of exceptions would be very extensive, but those most commonly existing are outlined below.

*1 Abnormal loads and amounts*
*a*    Peak amounts of documents caused by seasonal or other conditions.
*b*    Figures and totals greatly exceeding the normal, often associated with the above situation. This is particularly relevant to real-time systems because complex arrangements have to be made in order to deal with various degrees of overload. A priority struc-

ture must be established to cover the various types of messages so that the computer can decide which messages to give precedence to.

c    Essentially positive figures becoming negative, usually caused by the processing of transactions getting out of phase, e.g. 'negative' stock-in-hand caused by stock issues being processed before receipts.

## 2 Special considerations

a    In relation to new employees, unknown customers, introduction of new sales items, etc. — these need the implementation of special tasks before they are absorbed into the system.

b    Priorities given to certain customers or jobs — these priorities may lose their point with the introduction of a data processing system.

Allowances made outside those formulated, e.g. bonuses, prices, and discounts decided manually at the time and place of application (this arrangement may be common enough to be regarded as normal).

## 3 Periodic additional work

a    Extra work just prior to statutory holidays, especially for payroll staff, stocktaking and evaluation.

b    Financial year-end work.

c    Periodic and *ad hoc* reports for management.

## 4 Missing Data

a    Entries omitted from documents, particularly those originating externally, and cross-references missing from other records.

b    When the input to the computer is on-line, missing data is detected at an early stage in the system. Nevertheless the exact data required as obligatory input and the characteristics of optional data must be determined.

## 5 Mistakes

a    Non-agreement of control totals; what procedures exist for their reconciliation?

b    Arithmetic and transcription errors; how does their correction affect balancing and auditing?

c    Code numbers; are methods available for checking these?

d    As with missing data, an on-line system is able to detect certain mistakes in the input messages and reject or query these immediately. It is therefore important to ascertain what mistakes are likely to occur and how these are best corrected.

## 4.5   Human aspects of systems investigations

Mention of the word 'investigation' immediately engenders a feeling of apprehension in some members of staff. If followed by terms such as 'computer systems' or 'data processing', this apprehension rapidly develops into a state of acute anxiety. These conditions are really caused by two main factors: first, a fear of the unknown, and second, fear of the possible loss of status. The latter culminates in a genuine fear of being made redundant; this is especially prevalent amongst older clerical staff. The fear is, more often than not, without foundation since there is usually an over-all shortage of clerical staff within the organization. This means that they can be absorbed by other departments if made redundant by the data processing system, including employment as an operator, programmer, or control clerk within the data processing department itself.

The systems analyst can start to allay these fears during his interviews with the members of staff in the investigation stage. The 'unknown' factor may take several forms, and is generally the computer, the new system, new types of staff, or the company's future policy. A brief discussion of these points will put them in proportion, and although the systems analyst cannot be expected to provide all the answers at this stage, the clerical staff will feel happier in the knowledge that their interests are being given consideration.

A general explanation of the purpose of the investigation, as issued by management and reinforced by the systems analyst, is invaluable. A discussion of the impending changes gives the analyst an opportunity to establish a rapport with the departmental staff, and to encourage them in their desire to be involved in the new system. To suggest to staff that involvement with data processing will very probably increase their status and potential is in no way deceitful; a brief inspection of the 'situations vacant' columns quickly confirms this. The degree of interest shown by each member of staff is usefully noted for later consideration during the recruitment of data processing staff.

At the investigation stage, it is often impossible for the analyst to give accurate replies to direct questions about a particular person's future prospects. It is better for him to be non-committal than to administer 'soothing syrup' in the form of vague blanket assurances. These are apt to backfire on him when radical changes occur later and at the time when the utmost co-operation is needed from the staff. For further information on the sociological aspects of data processing the reader is directed to the further reading given in Section 4.9.

## 4.6  Interviewing

Interviewing forms an essential part of a system investigation, and should be thought of as discussions rather than as formal interrogation sessions. An interview that acquires the semblance of a formal eye-to-eye confrontation is doomed to failure. The systems analyst must always bear in mind that he is more the potential beneficiary from the interview than is the interviewee.

It is necessary during a systems investigation to interview the appropriate members of staff at all levels. The analyst's approach must therefore be flexible enough to cater for all these levels and to deal with all types of personalities and situations.

All interviews should be arranged in advance to take place at a time when there will be a minimum of interruptions. If during the course of an interview there is a continual stream of interruptions, it is wiser to conclude it prematurely than to allow it to disintegrate; it can then be restarted later under more favourable conditions. This also applies when the interviewee is under obvious stress, caused by either the interruptions or another reason such as a sudden rush of work.

The levels of the staff with whom contact is made during an investigation vary from one organization to another. A person's title conveys little meaning in itself, and can only be assessed in relation to the size and structure of the organization in which he works. Broadly we can consider there to be four levels of staff, and the most suitable approach is adopted accordingly. These levels, from the systems aspect, are :

> Top management (who decide policy).
> Line management (who control procedures).
> Skilled staff (usually senior clerical staff).
> Other staff (who follow a routine pattern of work).

*Top management interviewing*

Contact between the systems analyst and top management will have already been made during the assignment and feasibility stages. Following upon this initial contact, every effort must be made to obtain individual as well as collective views. What information should be looked for from this level? The systems analyst cannot merely adopt the role of questioner and thus obtain all he requires — he is as much concerned with hearing opinions as in obtaining straightforward answers. He should not be over-awed by top management, remembering that they are likely to be just as much in awe of new systems, computers, and suchlike.

The main points to be covered in interviews with top management

are as follows:
1   Take every opportunity to cultivate the seeds of top management's comprehension of data processing that were sown at the feasibility survey stage.
2   Obtain ideas, opinions and facts in relation to objectives, competition and major problems.
3   Avoid asking for trivial facts or precise details of methods.
4   Ascertain the future strategy of the firm as regards mergers, take-overs, marketing and manufacturing policies. Assurances should be given by the systems analyst that this information will be treated as confidential, but nevertheless it will not always be forthcoming.
5   Inquire about the structure and administration of the company, particularly as regards the relative positions and responsibilities of the line managers.
6   The employment of a tape-recorder during the interview might be acceptable at this level; permission should be obtained first however.

### Line management interviewing

The points to be included are as follows :
1   When making arrangements for the interview, give the manager some idea of its subject and purpose.
2   Before the interview takes place, brief oneself on the general duties and position of the interviewee, and also on the topics to be discussed. Vagueness on the part of the systems analyst results in the manager's loss of respect for him.
3   Try to involve only one other person at a time; if members of the manager's staff are called in to answer specific queries, get them to depart as soon as this has been done; alternatively postpone their intervention until the end of the interview. Do not ask the manager's staff, in his presence, questions that have already been put to him, even if he failed to provide an answer. These are better asked later when interviewing the staff privately.
4   Avoid questions concerning higher-level policy, but discuss the policies he formulates. Ask for suggested improvements to the existing system.
5   If the manager is likely to be continually interrupted, suggest that the interview is held elsewhere than in his office; for instance, in the analyst's office or in another quiet and private room.
6   Ask about the duties and responsibilities of each senior member of his staff, and make discrete inquiries regarding their person-

alities so that any difficult characters can be anticipated. Seek permission to interview his staff, explaining briefly the reason for this requirement.

7    Control the interview:
   — avoid wandering too far from the subject,
   — do not allow generalizations to obscure the actual situation,
   — separate opinions from facts

8    Conclude the interview with a quick résumé of the ground covered, ask if there are any outstanding points but leave an entrée for further contact. Doubtful facts can be confirmed later by sending a memorandum containing the systems analyst's interpretation of the ambiguous points

*Skilled staff interviewing*

With these people, the points to bear in mind are as follows:

1    Restrict the questions to details of his duties only, avoid discussing policy, do not pose leading questions.

2    Show a competent interest in his work, avoid condescension, and do not make adverse comments.

3    Do not allow his manager or other members of staff to be present all the time; discussion groups are more useful at a later date.

4    Take notes, explaining the reasons for these to the interviewee.

5    Allow time for him to collate any required lists and copies of documents, returning a few days later to collect these if necessary. It is better not to allow the interview to become broken up by searches for documents but instead to collect these at the end of the interview or later.

6.   Do not attempt to cover too much ground in one interview; one hour at a time is sufficient. If necessary make an appointment to continue the interview at a later date, preferably within a few days. Other points are the same as 4,7 and 8 of line management.

*Other staff interviewing*

It is not usually necessary to interview more than a small proportion of the other staff, as it is almost inevitable that the details of their work will have been already explained by the more senior staff.

The points to remember are, however:

1    Keep the questions simple, and do not encourage the stating of opinions.

2    Allow him to demonstrate his work rather than explaining it
     verbally, obtaining copies of the documents he uses.
3    Be friendly, adopting a neutral attitude towards his relation-
     ship with management if this subject arises.
4    Show interest in his work, avoiding condescension and adverse
     comments, and be complimentary whenever possible.

## 4.7    Staff organization and utilization

*Staff organization*

The over-all structure of the organization will, by this state, be
well known to the systems analyst. With this knowledge the relevant
departments can be chosen for further examination and their
structure put into the form of a Department Structure Chart similar
to the example shown in Figure 4.4. These charts not only help in
the planning of the investigation but are also used to prepare a Staff
Establishment Table, as described below.

A departmental structure chart has a box for each manager and
supervisor, and this is connected by lines to that of his superior, and
to the sections that he controls. A manager's or supervisor's box
contains his title, name, and possibly other information, such as his
location. Each section's box contains the name or duties of the
section, together with the number of staff in each 'grade'. The precise
meaning of grade as used on this chart cannot be universally defined,
but should be decided in relation to the size and type of company or
organization. In some large companies, and in local and national
government departments, staff grades are already defined; but otherwise the staff should be 'graded' unofficially by the systems analyst.
Three or four grades are sufficient and their allocation must be
confidential, since this is not in any way an attempt to compare
staff individually, but merely a means of obtaining an idea of the
structural balances of departments. It must be appreciated that in some
firms the preparation of this chart will not be a straightforward matter, particularly if the firm is organized according to a non-
hierarchical system.

The departmental structure charts are summarized and combined
to form a Staff Establishment Table as shown by the example in
Figure 4.5. This table is then used to provide a picture of the spread
of staff, and to check against the payroll distribution that all
personnel have been accounted for. It also indicates the areas for
potential staff economies and improvements in efficiency.

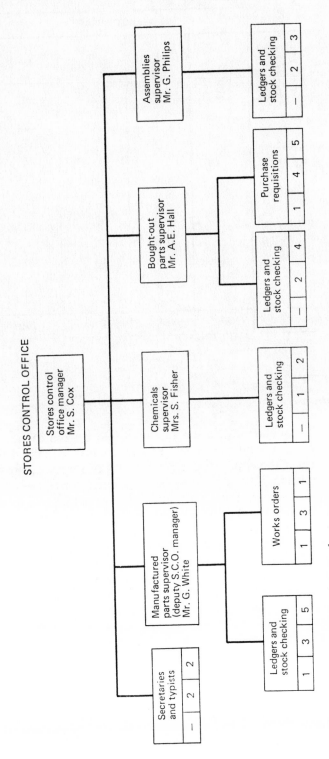

STORES CONTROL OFFICE

**Figure 4.4** *Department structure chart*

51

| | NUMBERS OF STAFF | | | |
|---|---|---|---|---|
| Grades / Departments | A | B | C | Totals |
| Stores control office | 8 | 17 | 22 | 47 |
| Purchasing | 4 | 9 | 13 | 26 |
| Production control | 10 | 25 | 20 | 55 |
| Wages | 6 | 10 | 9 | 25 |
| Sales invoicing | 5 | 9 | 18 | 32 |
| Totals | 58 | 98 | 114 | 270 |

**Figure 4.5** *Staff establishment table*

*Staff utilization*

A rough measure of staff utilization can be obtained from each departmental manager and cross-checked later against individual statements. These cover the approximate percentages of time spent by each person on his main tasks, and are subsequently converted into hundreds of hours per annum after allowing for overtime and special annual work. The resultant Staff Utilization Table as shown in Figure 4.6 is used to highlight workloads that might be susceptible to diminution.

Whilst gathering this information, the systems analyst should also inquire about other factors of interest such as those below.

1    What arrangements exist for coping with unexpected peaks of work, sickness, bank holidays and annual leave?

2    Are there any avoidable overlaps of staff activities, such as might, for instance, be connected with the transferring of data from one record to another?

3    What duties are currently performed that are completely outside the scope of a computer, for example, answering telephone enquiries from customers. These tasks, although sometimes occupying a negligible amount of time, can be of vital importance in the functioning of both the existing and the new system.

| STORES CONTROL OFFICE | STAFF UTILIZATION IN HUNDREDS OF HOURS PER ANNUM | | |
|---|---|---|---|
| Grades / Activity | A | B | C |
| Filing | – | 20 | 50 |
| Docket sorting | – | – | 52 |
| Ledger entering | 15 | 80 | 105 |
| Stock recording | 25 | 52 | 60 |
| Purchase requisitions | 40 | 68 | 58 |
| Works orders | 30 | 55 | 16 |
| Typing | – | 20 | 31 |
| Other work | 50 | 35 | 43 |
| Idle time | – | 10 | 25 |
| Totals | 160 | 340 | 440 |

**Figure 4.6**   *Staff utilization table*

4   The quantities and types of machines used, with their degree of utilization. These include keyboard accounting machines, adding machines, typewriters, punched card machines, copying machines and any other machines that process data.

5   The skills possessed by members of the staff outside their normal duties; these may have been obtained elsewhere and not officially recorded. They include the ability to operate office machines, a special aptitude with figures, a knowledge of particular systems and techniques, and especially any knowledge connected with computers, data processing and allied subjects.

## 4.8   Costing the existing system

Part of the systems investigation consists of estimating the costs incurred in running the existing system. The purpose of this exercise is to create a basis of comparison with the new system(s) that is subsequently designed. It would otherwise be difficult to justify the introduction of the new system, even allowing for intangible advantages.

The cost estimates of the present system must therefore cover all areas of work that are likely to be transferred to the new system. The costs need to be analysed between departments and activities, for the future as well as the present day. Thus it is necessary to predict the existing system's costs over the next five years or so. This is not normally too difficult provided no upheavels are expected in the organization.

In times of inflation it may appear that costs are increasing in an unpredictable manner. What we should remember however is that the new systems costs will inflate in much the same way. Whatever inflationary increases are applied should be applicable to both systems provided the same time span is covered. It is important not to confuse inflationary cost increases with real increases. The former costs are not meaningful in terms of effort and resource amounts. Real cost increases imply that more effort (labour and machine) is needed in the future to pay for them and they can be sensibly discounted back to the present day in order to compare two or more alternative systems (see Section 12.1).

Costs can be broken down into capital expenditure and recurrent expenditure. The former implies a once-only cost, such as the outright purchase of a piece of equipment, which is then depreciated over a number of years. The theory and practice of depreciation as an accounting method need not be of concern to the systems analyst as he is interested in actual cash flows. A capital expenditure can therefore be safely allocated to the year in which it causes the actual cash outflow.

Recurrent costs apply to rented or hired equipment, wages and consumable materials. There should be no difficulty in estimating future recurrent costs as long as nothing untoward happens in the labour market.

The extent to which cost information is readily available to the systems analyst during the investigation depends largely on the organization's costing/budgeting department. This information may, on the one hand, be available in detail; on the other hand, it may be supplied only in rudimentary form. In either case it is probable that the systems analyst will need to further analyse the cost information in order to bring it into a form such as in Figure 4.7.

The labour costs (wages) in the department costs table are approximated from the staff utilization table (Figure 4.6) by applying an average wage for each staff grade. This average is available from the wages office. It is not worth using the actual wages of individual staff for these computations as great accuracy is unnecessary.

It is the staff wages that dominate office costs but it is also necessary to apportion machine and consumable material costs between the various activities likely to be taken over by the new system. In practical terms it is not worth apportioning the capital cost of an

| STORES CONTROL OFFICE | COSTS £ PER ANNUM | | | | |
|---|---|---|---|---|---|
| Activity | Wages | Equip-ment | Stationery | Overheads | Totals |
| Ledgers entering | 36,250 | 3,750 | 500 | 10,000 | 50,500 |
| Stock recording | 26,900 | 3,100 | 300 | 7,500 | 37,800 |
| Purchase reqs | 34,300 | 5,700 | 450 | 10,000 | 50,450 |
| Works orders | 22,400 | 2,600 | 350 | 6,250 | 31,600 |
| Other work and idle time | 62,150 | 2,850 | 100 | 16,250 | 81,350 |
| Totals | 182,000 | 18,000 | 1,700 | 50,000 | 251,700 |

**Figure 4.7** *Activity costs table*

existing machine unless this is high. The money has already been spent and ceasing to use the machine does not bring a refund unless it can be sold for a significant sum.

Recurrent costs, such as rented equipment and consumable materials, are more likely to be apportionable between activities, and in some cases are known accurately from the cost records. Overheads, such as rent, rates, heating and insurance, need to apportioned if some of the accommodation will be released by the new system. If no established method of overhead apportionment exists, it is good enough for these purposes to apportion them in proportion to the floor area occupied by the department in question. Within this they are apportioned to activities in proportion to the sum total of the wages and machine rentals of the activity.

The costs table of Figure 4.7 shows certain costs derived from Figure 4.6 on the assumption that the average pay of grade A staff is £3 per hour, grade B £2 per hour and grade C £1.50 per hour, and that the department's overheads amount to £50,000 per annum.

## 4.9   References and further reading

4.1      MARTIN, *Design of real-time computer systems*, Prentice-Hall (1967).

4.2     ROTHERY and MULLALLY, *Practice of systems analysis*, Business Books (1972).

4.3     THIERAUF, *Data processing for business and management*, Part V, John Wiley (1972).

4.4     KELLY, *Computerized management information systems*, Chapter 7, Collier-Macmillan (1970).

4.5     FITZGERALD and FITZGERALD, *Fundamentals of systems analysis*, Chapters 1-5, John Wiley (1973).

4.6     BURCH and STRATER, *Information systems*, Part 1, Hamilton (1974).

4.7     DANIELS and YEATES (ed), *Basic training in systems analysis*, Chapters 1 and 2, Pitman (1971).

4.8     MUMFORD and WARD, *Computers, planning for people*, Batsford (1970).

4.9     WARNER and STONE, *The data bank society*, Allen & Unwin (1970).

4.10    'People problems', *Business Automation* (October 1970).

4.11    'Pawns in a magic game', *Data Processing* (July-August 1969).

4.12    HADLEY, *Elementary statistics*, Holden-Day (1969).

4.13    MARTIN, *Design of man-computer dialogues*, Prentice-Hall (1973).

4.14    *Data processing flowchart symbols*, BS 4058:1973 and ISO 2636-1973 (E), BSI (1973).

# Chapter Five
# *Analysis of basic data*

## 5.1   Classification of items

Many well-established office and factory systems function quite satisfactorily without using code numbers of any type. These systems are
usually completely manual and can be operated in this manner owing
to the staff's long-standing familiarity with the uncoded entities.
This situation can continue for many years without catastophe, but
a change of circumstances can quickly put an end to this happy state.
Among such changes are the employment of new staff, who would
not be familiar with the entities, a change or rapid increase in the set
of entities handled, or the introduction of new methods or machines.
When any of these events occur it may become necessary to allocate
code numbers to the uncoded entities, and before doing this, it is
worth considering the ways in which a set of entities could be classified. The method to be adopted depends primarily upon the subsequent uses to which the classification will be put. It is of no avail to
classify raw materials according to their colour for instance, if this is
of no significance in the use which will be made of the materials.

   The way in which a set of entities is classified will be  reflected in
the allocation of code numbers to the entities. The code numbers encompass the classification, and to some degree can contain more than
one classification, but we must be careful not to over-classify within
the code number or this will be too long and complicated.

*Changeable classifications*   A set of entities may be classified at one
point of time into what appears to be unchangeable groups. At a
later date a change of circumstances, perhaps completely outside the
control of the company, may make this classification absolutely value-

less. If it is known and accepted that this can happen, the system must be planned so that entities can be reclassified whenever necessary without at the same time disrupting the whole system. The most important point here is to maintain the continuity of identification of items and, with this in mind, an arbitrary identification code should be used for each entity in addition to the classifying code.

## Methods of classification

Whatever method of classification is decided upon, it must fulfill all the cardinal requirements of a classification system; these can be summarized into the three points below:
1   The method must provide for everything that is to be classified both now and in the future.
2   It must be clear and concise, with readily understood logic — the classification into which an entity falls must be immediately apparent.
3   It should not be more specific than is needed for the uses that will be made of it; over-comprehensive methods tend to engender long and complicated code numbers.

*Hierarchical classification*   When using this method, entities are classified into separate groups according to their most significant characteristic, and then sub-classified within these groups according to another characteristic, and so on. Thus within a hierarchical list, there is one place only for a particular entity.

Suppose, for example, it is required to classify a range of raw materials according to three of their main characteristics — substance, price bracker, and form. Using hierarchical classification, this could result in the following arrangement:

Class 1   VALUABLE MATERIALS
   *Sub-class*        1   Non-ferrous metals
                        1. Sheet
                        2. Bar
   *Sub-class*        2   Chemicals
                        1. In measured containers

Class 2   MODERATELY EXPENSIVE MATERIALS
   *Sub-class*        1   Non-ferrous metals
                        1. Sheet
                        2. Bar
   *Sub-class*        2   Ferrous metals
                        1. Sheet
                        2. Bar
   *Sub-class*        3   Chemicals
                        1. In measured containers
                        2. In bulk

| Sub-class | 4 | Timber |
| | | 1. Plywood |
| | | 2. Planks |

Class 3  INEXPENSIVE MATERIALS
| Sub-class | 1 | Ferrous metals |
| | | 1. Sheet |
| | | 2. Bar |
| Sub-class | 2 | Chemicals |
| | | 1. In measured containers |
| | | 2. In bulk |
| Sub-class | 3 | Timber |
| | | 1. Plywood |
| | | 2. Planks |

As can be seen from the above example, non-existent items are deliberately not catered for, e.g. there is no classification for valuable timbers. Using this method, gold bars would be classified as 112, teak planks as 242, and so on. In the above example, only 17 different combinations of characteristics are necessary out of the 48 that are theoretically possible.


*Faceted classification*   With this method each entity is coded in a way intended to define its characteristics (facets). There is normally one position in the code number for each facet. In contrast to hierarchical classification, the meaning of a facet is known immediately from the value of the appropriate position in the faceted classification. In other words, this meaning does not depend upon the value of any other position. Faceted classification codes tend to be longer than hierarchical codes because of this improved interpretability.

Thus, taking the same range of raw materials as above, faceted classification could result in the following codes:

Facet (A) SUBSTANCE
 1. Non-ferrous metals
 2. Ferrous metals
 3. Timber
 4. Chemicals

Facet (B) PRICE BRACKET
 1. Valuable
 2. Moderately expensive
 3. Inexpensive

Facet (C) FORM
 1. Sheet or plywood
 2. Bar or planks
 3. In bulk
 4. In measured containers

Thus a material such as gold bars would be 112, sheet steel 231, bagged cement 434 and so on. There is always some redundancy with faceted classification because certain combinations, although catered for, cannot exist, such as valuable ferrous metals, or timber in measured containers.

## 5.2    Coding of entities

The use of code numbers as a means of identifying things and persons has become well established within the spheres of business and production. With the growth in variety and complexity of manufactured articles, it has become impossible to identify each item in an unambiguous way except by giving it a unique code number. As a result many code systems have been introduced throughout the years, with various degrees of efficiency. The majority of these systems are adequate for data processing provided they are used correctly and consistently. Any code system already in use that is unsuitable for a data processing system should be replaced. The advantages of a new code system from the data processing viewpoint should be weighed against the disadvantage of changing over, with its inevitable problems and the errors that will at first result from it. Changes made to even a small set of code numbers can cause much confusion among its users, and it is advisable to employ both the old and the new code numbers together for some time before phasing out the old numbers.

Within a data processing system, code numbers are the lubricants of the computer routines; without them the system would sieze up owing to the computer's inability to identify the entities with absolute certainty.

*Advantages of code numbers*

What are the particular advantages to be gained from the employment of code numbers in data processing?

*Identification*    An entity can be identified with absolute precision, so that there is no possibility of two different things being confused. These requirements are even more important in a data processing system than in a manual system because there is less chance of the former recognizing an uncoded or erroneously coded entity. The computer need not be completely guileless in this respect, however, as can be seen in Section 5.3 and 5.4.

*Storage space*    A code number identifies an individual entity by using far less characters than is needed by a plain language description. Whereas a 3-digit code number uniquely identifies 1000 different items, these would need descriptions of up to eight alphabetic characters. This represents an increase in computer storage space of five times, and there is also more chance of a mis-identification occurring.

In addition to reducing the storage space required, code numbers also demand less effort for keyed and punched data entry.

*Access to stored data*   As is explained in Chapters 7 and 8, code numbers form the keys by which data records on computer files are recognized. They enable the programmer to instruct the computer so that it can gain access to individual records.

*Grouping of entities*   Similar but not necessarily identical entities are recognized as such by the computer. This facilitates the sorting, merging and summarization of data needed in preparing analyses and totals.

*Design of code numbers*

The classification of an entity is represented by a group of digits at the start of its code number, the remainder of the code number consisting of an arbitrary group of consecutive digits. There are however many possible arrangements of digits and characters that can be devised, and when selecting an arrangement the systems analyst should always bear in mind the problems of the subsequent users of the codes.

The most important points to bear in mind when designing code numbers are as follows:

*Brevity*   Code numbers should be as short as possible, particularly if they will be involved in frequent transcription from one document to another. The employment of alphabetic characters furthers this end; three alphabetic characters provide over 17 times the range of three numeric digits. Seven digits or characters are about the maximum that can be safely carried in the head. More than this entails the user in splitting the code number into two or more sections.

*Visual dissimilarity*   If it is advantageous to use alphabetic characters in a set of code numbers but the full alphabet is not required, the characters are best chosen according to their dissimilarity in appearance. This dissimilarity is, to some extent, a matter of opinion, but a suggested list is:

X W L Y F K M P N E A V H Z T R U B D J Q C S G I O

This list has the more dissimilar letters on its left. Thus, if only six letters were needed for a set of code numbers, they would be XWLYFK (see [5.7]).

*Audio dissimilarity*   If the code numbers are more likely to be quoted (in English) than written, the list in order of dissimilarity could be (omitting Z in the USA):

ZROWQHYILMNUKXJSFACTBDVGPE

If a limited number of characters suffice for a code set, those with similar sounds could be omitted. Thus although Y and I both appear towards the head of the list because they are both dissimilar to the other letters, one could be omitted to avoid confusion with the other.

*Upper case (capital) letters only*   The use of both upper and lower case letters within one code set leads to confusion. Since computers generally print only upper case, these are the better choice. Similarly symbols (asterisks, obliques, dashes, etc.) are to be avoided unless absolutely necessary for special reasons.

*Consistency of layout*   All the code numbers within a set, i.e. applying to one set of entities, should have similar layouts. This identity of layout encourages both accuracy and completeness in their use. For example, if a set of 700 code numbers is required, it is better to have A100 to A799 than A1 to 700. If it is particularly desirable to start at A1, then non-significant noughts are introduced, i.e. A001 to A700.

Different entity sets should preferably have dissimilar layouts in their code numbers. This minimizes the chance of confusing one set with another when they are used on the same document or in the same routine.

*Gaps within codes*   Gaps between the characters or digits in a code number are better avoided, since they tend to stray and thus cause confusion when keying the code number. When gaps are already in existence, they can be effectively eliminated within the computer by omitting them when keying; they must then however always be in the same position(s) in each code number set and can then be reinstated before the code number is printed or displayed.

*Redundancy*   As far as reasonably possible, all the positions in a code set should carry the maximum information. This applies particularly to the classification positions of a code number. Referring to the faceted classification example in Section 5.1, the 3-digit code allows for 729 ($9^3$) different classifications. In fact, only 48 (4 × 3 × 4) classifications are needed, this is a redundancy of over 93 per cent.

*Uniqueness*   In assigning code numbers to entities, the two cardinal rules are (*a*) the same code number must not be assigned to two different entities, and (*b*) the same entity must not end up with two different code numbers. These rules are easily contravened when large sets of entities have to be coded. The sorting and comparison capabilities of a computer are usefully employed in checking for code error.

*Code systems*

*Serial codes*   These are simply numbers or letters assigned to entities in an arbitrary manner. Usually the next available code number is assigned to an entity when it first makes its appearance. Serial codes carry no information about the entity but are straightforward to assign and have low redundancy. This method is suitable for a large set of entities that have no usable characteristics, e.g. personal customers of a bank.

*Sequential codes*   The entities are arranged into some significant order, e.g. alphabetically by name, and the code numbers are then assigned sequentially. This method has the advantage of obtaining a meaningful order in the entity set if the code numbers are arranged in sequence. It has the inherent drawback that if the numbers are assigned incrementally, new entities cannot generally be inserted into their correct places in the significant order. A way round this problem is either to leave gaps between successive code numbers or to employ partial sequential codes.

*Partial sequential codes*   The entities are arranged into an order that has a partial meaning, e.g. the first two letters of a set of surnames. The code numbers are assigned incrementally but with a gap between groups for the insertion of new entities. There is no point in arranging the entities into a precise order if new entities are to be inserted later since this will inevitably destroy the sequence.

*Block codes*   A block of consecutive code numbers is allocated to a general class or group of entities, allowance being made for expansion of the group. The code numbers have no particular significance as far as individual items are concerned except that they indicate which general class the item falls into. A typical example of this arrangement would be to give code numbers 1000 to 4600 to manufactured components, 4601 to 5500 to manufactured assemblies and 5501 to 5900 to manufactured products. A slightly more useful arrangement would be to assign the blocks so that the first digit of the code number indicated the group.
   Block coding has the advantages of brevity and simplicity but does not provide much meaning nor the facility for instant identification of the item concerned.

*Interpretive codes*   As the name suggests, these code  numbers can be interpreted from the values of their numerals or letters so as to provide a partial specification of the items. This method is useful for a set of entities that has a limited number of characteristics so it is then possible to incorporate them into the code number. An example of an interpretive code would be 1204, meaning a bar of cross-section 12 cm by 4 cm. Other candidates for interpretive coding are car tyres, fluorescent tubes and screws.

*Mnemonic or derivable codes*  When it is the case that codes have to be remembered by people without reference to manuals or lists, mnemonic coding may prove useful. Although it is unlikely that all of a large range can be remembered in this way, at least partial identification is possible from inspection of the code. A simple variant of this method is to incorporate part of the entity's name into its code, for instance in the UK postal codes, CV = Coventry, LS = Leeds.

   Wherever possible the most randomly distributed value positions of the names should be used so as to minimize the synonyms that occur. A derivable code is created from the entities' names according to strict rules but synonyms are difficult to avoid.

## 5.3   Feasibility checking

A well known saying in the data processing world is 'rubbish-in, rubbish-out'. This suggests that if there are errors in a system's input, errors will arise in its output. This will undoubtedly be the case if it is allowed to happen, and so as many input errors as possible must be detected and eliminated by the system.

   The methods for checking the preparation of input data are described in Chapter 10 but, over and above these, a series of feasibility checks can be employed to detect errors that may have originated far away from the data processing department. These types of error include, for example, mistakes made by a customer in his order. Feasibility checks can, of course, be carried out in manual systems; the main difficulties are the time taken and the extra staff needed. The computer is, after all, only doing sophisticated common-sense checks at high speed. Feasibility checks help to eliminate errors brought about by factors such as faulty clerical work, incorrect keying (although this is also detected by verifying), mistakenly identified batches of data, e.g. last week's instead of this week's data, wrong files (also checked by other means) and missing input data.

   The degree to which feasibility checks are employed is decided by balancing the level of safeguard needed on the one side, against the additional programming necessitated on the other, not forgetting that additional programming occupies both time and computer storage space.

The types of checks that can be employed, either alone or combined together, are described in the next few pages.

*Limit checks*

Limit checks may be applied to both the input data and the output results from the computer. Each data item, i.e. self-contained unit of data, of an input record is checked by the computer program to ensure that its value lies within certain pre-defined limits, a similar arrangement applying to output data items. These limits are the maximum and minimun values that the data item can normally reach. Examples are: (*a*) the hours worked in a week by an employee, as stated on his clock card, must lie between 20 and 60, (*b*) the week's earnings of a piece-worker, as calculated during the payroll routine, must lie between £30 and £100.

*Fragmented limits*

The data item's value is pre-defined as lying within one of a series of separate limits. The data item may, for example, be a code number that must lie within one of the ranges 1 to 11, 20 to 28, or 40 to 52. Thus, codes such as 8, 24 and 40 would be accepted, whereas 12, 33 and 53 would not. The series of limits can include zero and infinity; so that, for example, if a data item must lie outside the range 100 to 400, it would have two alternative sets of limits imposed on it, i.e. 0 to 99 or 401 to $\infty$.

*Combination checks*

A combination check can be applied after two or more data items have been combined in some way or other. This combining is often nothing more than adding or multiplying the data items together; the check is then applied to the result.

**Example**    The limits of purchase quantity are 1 to 1000, and the limits of purchase price are £1 to £90. A purchase order for 500 items at £80 each, although passing the limit checks on the individual data items, could be detected by applying a combination check with a maximum of, say, £1000, to the value of the purchase order.

The combination check is equivalent to a limit check on intermediate or final results, and is especially useful for detecting output fields that have become too large for the printing space allocated to them.

## Restricted value checks

These apply where a data item may have one of a short series of individual values and no others. This is really the same as a series of fragmented limits each of which has its maximum equal to its minimum.

A typical example is to check that the Stores number associated with each of the stock transactions is equal to either 1, 2, 5 or 8, because these are the only stores in existence.

## Format checks

A format check is the means of ensuring that all the requisite data items are present in a record. This is especially applicable to variable-length records composed of a variable number of data items or data elements. A count of these is made by the computer during input and checked against a pre-known count held in the record.

A format check should ratify that the data item or data elements is actually present, not merely that a space has been left for them.

A typical example of a format checking is to count the data elements (ordered items) present in a customer's order and to compare this with the total items at the bottom of the order.

## Picture checks

The layout of each data item is checked for compliance with its proper 'picture' (Section 4.2). Every position must be one of the digits, letters or symbols expected and specified by the data item's picture. A simple picture check merely ensures that the position is one of the three types just mentioned. A more sophisticated picture check looks for absolute values, such as in the examples below.

**Example**  A part number consists of two alphabetic characters followed by three numeric digits, the acceptable values of which are:

| | |
|---|---|
| 1st position | A, B, C, D, G or M |
| 2nd position | A to M, or T to W |
| 3rd position | 1 to 6 |
| 4th position | 0 to 9 |
| 5th position | 0 to 9 |

Thus part numbers DF293 and MV605 are acceptable, but DF793, EA538 and AN123 are not.

## Compatibility checks

Two different data items or the results of two different calculations are checked for compatibility by applying a table of limits or restricted values. This check is best explained by means of an example — results of practical experience in a factory show that the total machine set-up time of a batch of products is roughly related to the batch quantity. Since both of these are calculated from data in the workers' job dockets, they can be checked for compatibility by using a table such as that below:

| BATCH QUANTITY | TOTAL MACHINE SET-UP TIME |
|----------------|---------------------------|
| 100-500 | 20-30 min |
| 501-2000 | 30-70 min |
| 2001-7000 | 60-160 min |

Thus a total machine set-up time of 57 min and a batch quantity of 450 both calculated from the same docket would not be acceptable.

## Probability checks

In the event of failure of any of the checks described in the preceding paragraphs, the computer would obey an error program and reject all or some of the related data. There are cases however where exact limits are not applicable, but where there is nevertheless a good

| COMMODITY GROUPS | CLASS OF TRADE | | | |
|------------------|---------|-------------|-------------|------------------|
| | 1 | 2 | 3 | 4 |
| | Stores | Large shops | Small shops | Clubs and firms |
| CODES | | | | |
| 10 — 29 heaters, irons | 30 | 10 | 3 | 2 |
| 30 — 59 radios, record players | 10 | 3 | 2 | 1 |
| 60 — 89 television sets, washers, cookers | 4 | 1 | 1 | 1 |
| 90 — 99 lamps, batteries, equipment | 1000 | 100 | 50 | 100 |

**Figure 5.1** *Table of probability checks*

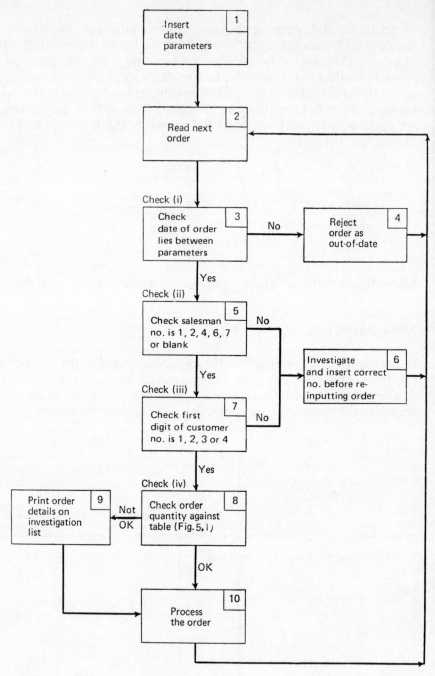

**Figure 5.2** *Flowchart of feasibility checks*

chance that values outside the limits are erroneous. These probable, but not absolutely certain, errors can be usefully reported by the computer for subsequent human investigation without being rejected at the point of detection.

The example shown in Figure 5.1 is drawn from the wholesaling of electrical goods, and shows the maximum order quantities to be expected in a single order for a commodity, as related to the commodity group and the class of trade. The probability check would cause the computer to highlight excess quantities for human investigation before the goods are actually dispatched.

## Parameters of feasibility checks

Owing to changing circumstances, it is quite possible that the limits imposed by a feasibility check at one date will no longer apply at a later date. This situation is catered for by treating the limits as parameters, i.e. as things that have to be assigned values prior to each time they are used.

A typical example is when checking the dates on sales tickets during a weekly sales analysis run. Since the dates should always relate to the previous week, the parameters are the dates of the first and last days of the previous week. These dates would be read into the computer just before the sales analysis run.

## Flowchart of feasibility checks

This is shown in Figure 5.2 and demonstrates how four types of feasibility check can be applied to an order received.

The details of an order record are:

| FIELD | LIMITS |
| --- | --- |
| Customer no. | First digit is 1, 2, 3 or 4 |
| Date of order | As per date parameters |
| Salesman no. | 1, 2, 4, 6, 7 or blank |
| Commodity code | None |
| Order quantity | As per table (Figure 5.1) |

The feasibility checks applied to the order are:
1   A Limit check on the date of order, using parameters (boxes 1, 3 and 4).
2   A Restricted Value check on salesman number (boxes 5 and 6).
3   A partial Layout check on customer number (boxes 7 and 6).
4   A Probability check applied to order quantity, using the table in Figure 5.1 (boxes 8 and 9).

## 5.4 Check digits

Check digits are used as a means of detecting errors that sometimes occur when numbers are transcribed from one document or medium to another. It is well known that the human handling of long numbers often leads to confusion with consequent errors appearing in the numbers. A prime example of this is in the dialling of long telephone numbers, where there is a high level of mis-dialling.

It has been shown, as a result of various tests, that errors of this nature fall into three main categories:

1    *Transcription errors*    Mistakes in copying a digit of a number, e.g. 13795 is copied as 18795; this category accounts for 86 per cent of all errors.

2    *Transposition errors*    Mistakes caused by swopping adjacent digits of a number, e.g. 42691 becomes 46291; these account for 8 per cent of the total errors.

3    *Other errors*    This includes the less common errors such as shift, double transposition, omission and insertion, and accounts for the remaining 6 per cent of the total.

A method of detecting the errors described above is to use an additional digit (or digits) along with the number itself; this is known as the 'check digit(s)'. This digit carries no meaningful information other than the assurance of the number's correctness. Its value is related to the rest of the number in such a way that any change in the number is reflected by a change in the check digit.

*Modulus II*

In order to obtain a high degree of security many check-digit methods have been devised. Most of these use a technique involving 'weights' and a 'modulus'. Each digit of the number is weighted, i.e. multiplied by a weight, the results are added together and their sum divided by the modulus. The remainder from the division is then subtracted from the modulus to obtain the check digit. Because the digits are weighted differently, transcription and transposition errors have a much higher chance of being detected.

A wide variety of weights and moduli have been suggested, and supported by statistical arguments [5.8 to 5.14]. The most prominent system is known as 'Modulus 11'; this uses a set of consecutive weights and a modulus equal to eleven.

The check digit is created as follows:

*Step 1*    Multiply each digit by its weight; the digit in the least significant position has a weight of 2, the next position a weight of 3, and so on.

*Step 2*   Add together the products of these multiplications.
*Step 3*   Divide this sum by 11, preserving the remainder.
*Step 4*   Subtract the remainder from 11; the result is the check
          digit. If remainder is zero, check digit is also zero. A check
          digit of 10 is normally written as X.

**Example**   The code 75264 would be operated upon as under:

*Step 1*   CODE NUMBER   7   5   2   6   4
          WEIGHTS               6   5   4   3   2
          PRODUCTS            42  25   8  18   8
*Step 2*   42 + 25 + 8 + 18 + 8 = 101
*Step 3*   101 ÷ 11 = 9 and 2 remainder
*Step 4*   Check digit is 11 − 2 = 9
          Code number is written as 752649

When a number is to be checked for validity, the check digit is
allocated a weight of 1; steps 1, 2 and 3 are then performed; the
remainder should then be zero.

**Example**   The code number is now 752649.

*Step 1*   CODE NUMBER   7   5   2   6   4   9
          WEIGHTS               6   5   4   3   2   1
          PRODUCTS            42  25   8  18   8   9
*Step 2*   42 + 25 + 8 + 18 + 8 + 9 = 110
*Step 3*   110 ÷ 11 = 10 and 0 remainder
          The code number is therefore valid.

A well-known use of modulus 11 is in international standard book
numbers as can be seen in the ISBN number at the front of this book.


*Other factors*

It is possible to use other values for the weights, such as 16, 8, 4, 2, 1
or various arrangements of consecutive numbers, e.g. 9, 10, 7, 8, 6, 3,
5, 2, 1 [5.11]. The modulus can also take other values and these are
often the prime numbers just below a hundred; in the case of 97, for
instance, there would be two check digits within the range 00 to 96,
giving a higher level of security than a single digit.
    If a modulus of 23 were employed, this would give a range of
check digits from 00 to 22. These could be conveniently written as
A to W, or as A to Z, if G, I and O were omitted (Section 5.2).

## Desirability of check digits

It must be remembered that by associating a check digit with a number, the over-all length of the number is increased. This increase will raise the various error rates in addition to making more work in writing the number. The advantages to be gained from the employment of check digits outweigh these disadvantages in the case of long numbers, but for numbers of less than five digits the advantages are marginal.

Another factor to be taken into consideration is the actual means of generating and verifying the check digits in a business system.

There are three main arrangements available:

1   Verification by a computer of check digits on source documents fed into the computer via optical character or optical mark recognition (Section 10.5).
2   Verification of check digits on source documents by a key-to-disk or similar system.
3   Generation of check digits from numbers on source documents by key-to-disk or similar system.


## 5.5   Interpretation of facts and procedures

As a consequence of the systems investigation a large amount of information is gathered together about the existing system and the situation within the organisation. Before utilising this information in designing a new system, it is wise to give consideration as to its true meaning and applicability. This may result in the need to gather further information or look for deeper explanations behind the facts. It is inevitable that these needs are directed towards the design of the new system since the investigation has not been carried out merely as an intellectual exercise.

The interpretation of the information gathered depends very much upon the particular company's circumstances and upon the applications investigated. There are no absolute rules applicable to all situations but a number of guidelines follow.


## Standardization

Before proposing any form of standardization we must give regard to its purpose and advantages. Standardization for its own sake is purposeless, and so what data, information and procedures can be usefully standardized within most organizations? Since the aims of standardization are simplification, error minimization and cost reduction, there are obviously many aspects that can be considered.

One of the things to avoid is giving the impression of unnecessary bureaucracy, and so the standardized procedures, documents, etc., should be self-explanatory and straightforward.

Examples of useful standardizations are:

1   Documents of a consistent size, layout and annotation.
2   Machines that are compatible so that work can be spread between them, e.g. typewriters with similar typefaces, photocopiers with the same capabilities.
2   Data items expressed in a consistent way, e.g. names, descriptions, addresses, dates, times and quantities.
4   Code numbers (see Section 5.1).
5   Identification methods, e.g. colour codes for document copies, abbreviations and symbols to represent actions and situations such as O/S for out-of-stock.
6   International, national, group and trade standards including customary ways of referring to things, e.g. a price expressed always in pence per dozen.

*Data items pertaining to entity sets*

1   Are the data items relating to an entity complete, bearing in mind all the likely processes that the entity will undergo?

2   Are the meanings of all data items understood in the full sense? Simple data item names such as 'labour cost' can be misleading.

3   Is it probable that new data items will appear in the future about which little is known at present, e.g. a new deduction against wages?

4   Which data items are best held in master records as against being computed when needed, e.g. VAT (sales tax) for an item of fixed price could be held in the master record to save calculating it every time a sale is made; alternatively, holding the item's VAT rate makes for easier amendment if the rate changes?

5   What interconnections exist between data items in different entity sets and what account should be taken of these? For instance, the materials in a set of manufactured products are connected with the supplier companies in the purchase accounts ledger.

6   Is there a high proportion of common values of similar data items in an entity set? If this occurs, it is generally possible to replace the data item value by a pointer, thus saving storage space. The pointer indicates the data item's value in a look-up table.

*Records in logical files*

*1 Data stability* The entities in a set each have a record in a logical file. To what extent are the entities likely to expand or contract in number over the foreseeable future? Whereas a database would handle this situation automatically, a non-database system could be strained by a large-scale increase in the number of entities. Trend analysis is often worth using in order to detect and predict a steady expansion of a file.

*2 Record identification* Changes to the pattern of entities' keys may cause inefficiency in the access method to the records. Also of interest are alternative keys (Section 6.2) that might be used for accessing the records. This information is of use in deciding the modes of storage and access of the files (Section 6.4).

*3 Record variability* The amount of data needed to be held in a record may vary from one entity to the next. To what extent is this unavoidable? Do the entities in a given set have data that varies for the same entity from time to time? The degree of data variation needs to be analysed in order for the systems analyst to decide on the type (fixed or variable length) of records to use (Section 6.1).

*4 Sequence* The need to hold or bring records into a particular sequence should be given consideration. The main reasons for sequencing are:
*a* To facilitate making changes to a logical file, i.e. inserting new records, deleting obsolete records, altering permanent data items, and inserting or removing changeable data items (Section 6.3).
*b* To derive summarised figures (counts and totals) for groups of entities.
*c* To combine or compare data items in records with identical keys. This may entail creating larger records from several small ones, or merely checking that the data items are in some way related.
*d* To highlight records containing data items whose values are in some way outstanding, e.g. excessive costs, low stock. This means that the records have to be sorted according to the relevant data item so that outstanding values are brought to one end of the sequence, i.e. high or low (Section 8.5).

*Errors and omissions*

Errors and omissions are often more easily detected than corrected. Nevertheless they need to be corrected at some stage and so we must consider precisely what this involves. The points at issue are:

1    If an error is detected in a batch of source documents, should
     the batch be rejected or held until the corrections arrive?
2    Can corrections be sensibly made at late stages of a routine?
     This could involve a manual correction made to computer out-
     put. Before embarking on this method it is vitally important
     that the systems analyst fully understands all its implications.
3    How are errors best reported? Frequent and unavoidable errors,
     such as casual customers' orders, have to be handled as very
     much part of normal procedures. Infrequent and unexpected
     errors must have a proper method of report. The main points
     here are who should be informed and how can we be certain
     that the correction has been made correctly?
4    Slight errors and omissions can occasionally be ignored. Obvious-
     ly this depends entirely on their extent and the effects on out-
     put. For example, a few sales tickets omitted from a global sales
     analysis are hardly likely to matter. On the other hand the
     equivalent cash missing from the till calls for investigation.


*Source data*

Source data is frequently one of the less controllable aspects of a
system. This is often the case because it comes from external
agencies, such as supplies and customers, or from dispersed locations
within the organization. Particular points of interest are:

*1  Reception*   Documents holding source data are created in a
number of ways, such as receipt through the mail, by phone or telex,
or in the organization itself. Whatever method is involved, it needs
to be studied so that a comprehensive control system can be devised.
The two main factors are *(a)* to seal all loop-holes against loss of
source data prior to its arrival in the data processing department and
*(b)* to ensure that the data is routed directly and speedily.

*2  Preparation*   Source data normally needs to be converted into a
form acceptable to a computer. This is perhaps the most exacting
and labour-intensive activity in a data processing system once it
becomes operational. The systems analyst ought to study all existing
documents carefully with a view to considering their eligibility as
source documents. Is the layout reasonable, are the entries legible,
are they of a sensible size, are the entries dispersed amongst many
other irrelevant data, and so on?

*3  Fluctuating volumes*   Why does source data vary in amount
from day to day? If this is the situation, it may be worth investiga-

ting the reason in order to smooth the arrival pattern and so obtain a
better work balance.

*Information output*

The information required by a manager may already be in existence
and in need of no change. In contrast, he may request completely
new information. Whichever is the case, the systems analyst is advised to give thought as to the genuineness of the purported need.
It is not an attempt to prevent managers from obtaining information
but simply to avoid waste of effort in preparing it. Having verified
the need, the principal aspects are:

*1   How is the information to be presented?*   The recipient will
have ideas in this respect but it is also worth suggesting standard
documents and layouts and also introducing the idea of non-
documentary information such as visual display and microfilm.

*2   To what use is the information put?*   The answers to this
question are the determining factors as regards layout, sequence,
number of copies and content. By knowing the usage of information, it is often possible to organise it so that the recipient can use
it quickly and efficiently.

*3   Will requested information still be needed when the new system
becomes operational?*   One must be on guard against accepting the
need for information regardless of future intentions. Information
that is of importance in the present system may become redundant
when a new system takes over.

*User department staff*

During the systems investigation the analyst has met many of the
present staff of the user departments and now is the time to think
about their future job prospects. It may be that some of the staff
will become redundant; others will be invaluable in the new system.
The decisions in this respect are not entirely those of the systems
analyst. It is his responsibility to make suggestions and that of
management to take the final decisions bearing in mind current
legislation.

Having decided which members of staff are to be retrained, their
retraining calls for consideration. It should be apparent for which
jobs they are most suitable, and so the first stage is to discuss this
with them. This is primarily the responsibility of the personnel
manager advised by the systems analyst.

*Existing system*

In some situations the existing system can be successfully translated into a new system without altering its inputs and outputs. Thus the user staff would not be aware of the change apart from perhaps slight changes to the output documents. The question arises as to the desirability of 'photocopying'. In view of the large amount of work necessitated by the changeover, it is obviously wise to give considerable thought to the improvements that can be introduced.

Another aspect of the existing system is whether its logic is applicable on a computer-based system. The systems analyst has to consider the extent to which the present decisions are convertable into computer-programmable form. Intuitive decisions and subjective judgements are not amenable to computerisation unless converted into quantifiable or logical form. This often applies to 'sharp-end' situations where decisions regarding actions, prices, allowances, etc., have to be made immediately. If these decisions cannot be formulated, then the new system must accept them as part of its input. This is not in any way disastrous so long as the situation is understood and the new system designed accordingly.

A further consideration is the activity cycle of the existing system. Routines may, for instance, have been carried out on a weekly cycle for time immemorial, but the advent of a new system introduces the possibility that a change would be advantageous.

There are few guidelines in this respect. Some routines are better cycled more frequently in order to improve cash flow or maintain better control over production. Others call for less frequent cycling in order to reduce the work by less setting-up and other economies of scale. Again, it should be remembered that the present cycle of an activity may be geared to circumstances no longer prevailing in a new system.

## 5.6    References and further reading

*Classification and coding*

5.1   'The construction of hierarchic and non-hierarchic classifications', *Computer Journal* (August 1968).
5.2   'Problems in constructing data processing codes', *Computer Bulletin* (June 1962).
5.3   *Commodity coding*, NCC (1968).
5.4   *Commodity information systems*, NCC (1970).
5.5   'An information measure for classification', *Computer Journal* (August 1968).

5.6 'Coding conflicts in the construction industry', *Data Systems* (January 1969).

5.7 'Presentation of alphameric characters for information planning', *Computer Bulletin* (November 1970) and *Comm. ACM* (December 1969).


*Check digits*

5.8 'A variant of modulus-11 checking', *Computer Bulletin* (August 1970).

5.9 'Modulus-11 check digit systems', *ibid.* (August 1970).

5.10 'Check digit verification', *Data Processing* (July 1966).

5.11 'An optimum system with modulus-11', *Computer Bulletin* (December 1967).

5.12 'The theory of modulus-N check digit systems', *Computer Bulletin* (December 1968).

5.13 'A modulus-11 check digit system for a given system of codes', *ibid.* (January 1970).

5.14 'Some error correcting codes', *Computer Journal* (February 1974).

## Chapter Six
# *Data processing files*

### 6.1   Purposes of files in data processing

The framework of a data processing system is its files; it is through the proper planning and control of the database that the system can function efficiently and comprehensively. The term 'file' as used in data processing has a more precise meaning than in manual systems. A data processing file is a collection of data in the form of 'records'. Each record is discrete and is labelled by a 'key' that is held as an identifier within the record. The various arrangements of files, records and keys are legion, and are devised to suit the requirements of a particular system. A set of interrelated files are, in effect, the database of a system. As explained in Section 6.7, a database is really more than this, but from the systems analyst aspect it can be regarded as the logical records and files necessary to provide the data required for processing other data.

What are the main purposes of data processing files?
1    To hold data in a form that enables it to be processed rapidly by the computer.
2    To make records accessible to the computer, either individually or *en masse*, and with|the degree of immediacy necessitated by the routines using them.
3    To provide security for the records, against both technical and felonious loss or damage.
These purposes are all achievable provided suitable hardware is made available for the system, and its files are employed in a well-planned manner.

Virtually any data processed by a computer is held on a file at some time or other. This may be either a temporary or a permanent state depending on the type of data and its purpose. An example of temporary data is the details appertaining to the issues of items from the factory stores; this data is valuable up to the point at which it amends a stock level or other information, thereafter becoming increasingly valueless. An example of permanent data is the stock levels of items held in the stores; although this data is continually changing, it is held permanently in the file for as long as the items are used in the factory.

In some applications the distinction between temporary and permanent data is less obvious, but in general terms, temporary data applies to transactions and activities whereas permanent data applies to reasonably stable facts.

## 6.2   File records

A data processing file, of whatever type, consists of a relatively large number of records — normally between 100 and 100,000. Each record consists of a number of data items, one or more of which act as the 'key' by which the record is identified. The data item(s) that comprises the key may not always be the same within a given set of records from one time to another. The computer is programmed to select the appropriate data items to use as the key during a particular computer run.

An example of a fairly simple record is shown below; there could well be one such record in a file for each item sold over a period of time.

| Data item | A | CUSTOMER NUMBER |
| " " | B | ORDER NUMBER |
| " " | C | WEEK NUMBER |
| " " | D | PRODUCT CODE |
| " " | E | QUANTITY SOLD |
| " " | F | SALES VALUE |

Data items A to D are the most likely to be the record's key. The computer would select the data item(s) forming the key in a processing run. In the above example none of the four data items provides unique identification of a record because their values are common to several different records. The exception to this could be 'order number' if this was unique to each item sold.

## Sorting keys

As well as being used for identification purposes, a key is also employed as a means of sorting records into a given sequence. These uses are of course the same thing, because the computer must identify a record before it can sort it. There is, however, a further significance attached to data items when used as sorting keys; this is their relative position when combined into one key. Thus, referring to the above record, if we wish to sort a file into customer number within product code sequence, the sorting key would be fields D and A in that order. If these data items were combined with A before D, the sort would be into product code within customer number. As a matter of interest there are 64 different ways in which four data items can be selected and combined to form a sorting key, but in practice many of these would never be used.

## Fixed and variable length records

The records in a given file are usually of the same type, i.e. contain the same types and numbers of data items, as in the above example. These are known as 'fixed length' records because each one occupies exactly the same amount of storage space in the file. There are, however, occasions when it is more convenient to employ records that consist of a variable number of data items. A typical example of a 'variable length' record is one that is formed from a customer's complete order, i.e. containing details of all the products ordered on the one occasion. Since customers do not all order the same number of different products, the record lengths vary between orders.

Using the same data items as in the previous example, the record now contains:

|  | | DATA ITEM | | | |
|---|---|---|---|---|---|
|  |  | A | CUSTOMER NUMBER | FIXED LENGTH DATA | |
|  |  | B | ORDER NUMBER | | |
|  |  | C | WEEK NUMBER | | |
| DATA ELEMENT |  | D1 | PRODUCT CODE | FOR FIRST PRODUCT ORDERED | VARIABLE LENGTH DATA |
|  |  | E1 | QUANTITY SOLD | | |
|  |  | F1 | SALES VALUE | | |
| DATA ELEMENT |  | D2 | PRODUCT CODE | FOR SECOND PRODUCT ORDERED AND SO ON | |
|  |  | E2 | QUANTITY SOLD | | |
|  |  | F2 | SALES VALUE | | |

Another cause of variable length records is that they contain variable length data items. The most common example of this is in name and address files. The varying lengths of names, streets and towns, combined with the differing numbers of lines in addresses, result in a variation from about 25 up to 125 characters per address. It is necessary in this case to insert 'end of field' markers between the name, street, town, and county so that the computer can split them up for printing in the conventional manner.

Although variable length data items may also apply to things such as descriptions of products, it is usually better to treat these as fixed length. This is done by allowing space in every record for the longest description, leaving blank spaces when shorter descriptions apply.

These arrangements can result in a file of records whose lengths vary considerably. In order that a computer can cope with them, it must be given some means of determining each record's length. This is achieved by the following alternative methods.

*Record length data item*    This is a data item in the record that specifies the length of the record. This is inserted when the record is created, and amended whenever the record changes in length. This can happen when further data items are inserted or if variable length data items expand. The 'record length' must be situated in the fixed length data of the record so that it can be located; often it is the first data item in a record.

*Data element count*    This is a data item whose value is the number of data elements (products sold in the above example) contained in the record. It is otherwise similar to a record length data item.

*End of record marker*    This is a special symbol or a group of digits or characters that do not occur elsewhere in the record, and is positioned immediately after the final data item in a record.

If variable length records need to be sorted according to a key in the fixed length data, no problem arises. To sort on a key in the variable length data however, the records have to be split up into sub-records of fixed length. It will be appreciated that otherwise the resultant sequence would be meaningless. When splitting records in this manner, it is necessary to replicate some of the fixed length data in each sub-record for use as keys.

An alternative replication is to 'chain' the sub-records by means of a 'pointer' in each one. This gives the location of the next sub-record in the chain, and so facilitates bringing the sub-records to-gether for sorting or amalgamation [6.1].

*Split records and chaining*

If either fixed or variable length records turn out to be extremely long, or if there are few variable length records that are much longer than the majority, the processing and storage allocation can become inefficient. It is advisable to fix a maximum length beyond which records are split into two or more smaller records. When this is done, the resultant smaller records are chained together by means of pointers, as above, and also keys may be replicated.

## 6.3 Classification of files according to contents

Within this heading the contents of a file can be considered as falling into one of three categories — transaction data, transition data, master data.

### Transaction (movement) data files

These files hold records that have been created from source data. The records normally contain all the data from the source and no other data. A transaction file may be sorted into a different sequence from its original, but still remains a transaction file provided the contents are unchanged. There is often an unpredictable total of records in the file, and a variable number, including none, having a particular key. For instance, a file of job dockets will vary in size from week to week and each man may have a different number of dockets corresponding to the jobs on which he has worked.

The contents of transaction files are not retained for long, but are replaced periodically by further transaction data. It is always possible to reproduce the source data from a transaction file by making a straightforward printed copy, after re-sorting if necessary.

### Transition data files

These files generally stem from transaction files as a result of changes made to the contents of the latter. The insertion or deletion of data items causes a transaction file to become a transition file, the insertions usually being derived from master files. A typical example of this process is the insertion of pay rates into a job docket (transaction) file from a pay rates (master) file.

Transition files are not retained permanently and are inclined to be even less permanent than transaction files.

### Master data files

Master files form a permanent feature of a data processing system, and hold records appertaining to diverse sets of entities. An individual entity set is usually semi-static as regards both its size and contents, and there is one record for each entity in the set. The records may be of fixed or variable length — typical master files with fixed length records are payroll records, commodity prices, stock levels, personnel records; and with variable length records are sales ledgers, purchase ledgers, parts operations, product and assembly contents.

A characteristic of master files, which is not applicable to transaction and transition files, is that during their life they are frequently 'updated', 'amended' and 'referenced'.

*Updating*   By this is meant the regular attention to each record in a master file in order to maintain it in an up-to-date condition. This is normally carried out at regular intervals and it is expected that a substantial proportion of records are updated on each occasion. For example, the monthly updating of a sales ledger file by invoice and payments data, or the weekly updating of a payroll file by earnings and tax data.

*Amendment (maintenance)*   This refers to the addition of new records, the deletion of obsolete records, and changes made to existing records that are not expected to occur regularly. An instance of amendment is the changes made to a price file when new commodities are introduced or when price adjustments are made to existing commodities. Maintenance is virtually the same thing as amendment but also includes the general tidying up (housekeeping) of a file.

*Referencing*   This is simply looking-up a record so as to obtain data from it without changing it in any way (input mode). An example is looking-up the prices of commodities during a sales invoicing run.

*Activity or hit rate*   This is a measure of the proportion of master records affected by an updating run, or referred to in a referencing run.

*Volatility*   This is a measure of the amount of amendment to which a master file is subjected. It is, however, a somewhat vague term and liable to be confused with activity.

Various other activity ratios can be applied to master files (see [7.11]).

## 6.4    Modes of file storage and access

Mode of access is liable to be confused with mode of storage because the same terminology is employed for both concepts. In some ways, the common terminology is acceptable since the two modes are closely connected in many cases. Mode of storage implies the way in which the records are organized in the storage medium, i.e. their sequence (if any), interrelation and structuring.

Mode of access refers to the method of examining or changing the records in a file. It is, essentially, the same as the sequence of the transactions that access the records. In general, the highest efficiency is achieved when the transaction records and the master records are in the same sequence.

Figure 6.1 shows the relationship between these two modes.

| Mode of access \ Mode of storage | Serial devices | | | Direct access devices | | | | |
|---|---|---|---|---|---|---|---|---|
| | Ser. | Part'd | Seq'l | Ser. | Part'd | Seq'l | Index-Seq'l | Random |
| Serial | ✓ | ✓ | ✓ | ✓ | ✓ | ✓ | ✓ | ✓ |
| Partitioned | | (✓) | | | ✓ | | | |
| Sequential | | | ✓ | | | ✓ | ✓ | |
| Selective-sequential | | | | | | | ✓ | |
| Random | | | | | | | | ✓ |

Figure 6.1    *Relationships of modes of storage and access*

*Serial storage and access*

A serial storage file is one in which the records are stored in adjacent locations so that the file is tightly packed and has no semblance of sequence or other organization. Typically, a batch of unsorted stock transactions is a serial file.

In order to find a particular record in a serial file, the search must start at the beginning and examine each record in turn. Thus, on average, half the records would need examining for each access.

Master files are not normally stored in serial mode.

*Partitioned storage and access*

A partitioned file consists of several partitions (members), each of which has a unique name or number, and within which the records are stored serially. It is used mainly for the storage of programs and sub-routines, or as a core dump (Section 8.3). Partitioned access implies that the computer proceeds directly to the start of a partition and then deals with the records in serial access mode. Although magnetic tape can hold a partitioned file, e.g. a library of programs, partitioned access to it is not truly possible because all the preceding data has to be examined in order to find the wanted partition. This is less true if the partitioned file is held on several reels of tape since the search commences at the start of the reel known to be holding the wanted partition. Partitioned access is quite straightforward on a direct access storage device holding a partitioned file; the records are simply examined from the start of the relevant partition onwards.

## Sequential storage and access

A sequentially stored file holds its records in some logical order based on the records' keys. This means that the records are stored in locations that have, in effect, successive addresses in the storage medium. Although magnetic tape is not addressable in the same way as a direct access device, the records when transferred from it are stored in contiguous locations in the main store buffer.

Sequential access means that the file records are dealt with in turn, and the search for a particular record proceeds onwards from the previous record accessed. This is obviously more efficient than serial access but the viability of sequential access depends on the hit rate (activity) of the stored records. If this is low, a long search is necessary to find even a few records, especially if this entails many transfers from a storage device with a long access time.

Where an inactive file consists of relatively few large records, the of the methods described in Chapter 7) and the record's location in store. The key of a wanted record is first found in the index (by one of the methods described in chapter 7) and the record's location in backing storage is deduced from this. This means that the records can be accessed fairly directly in backing store. This method, sometimes termed 'skip-sequential', does not really demand that the records are held in sequence, although they usually are. It is expensive of main store.

## Indexed-sequential storage and selective-sequential access

In order to save time when searching for a record, one or more indexes are created to enable the search to start nearer the wanted record's location. The search continues onwards from the location indicated by the index(es).

The corresponding mode of access is selective-sequential. This implies that the transactions may be arranged either serially or sequentially. The difference in efficiency between these two access modes is marginal if the file is finely indexed.

*Random storage and access*

Care should be taken not to confuse random access and random storage with direct access storage. The latter refers to a hardware device whereas the first two terms apply to logical concepts.

A random file is characterized by some predictable relationship between the key of a record and the location of that record in the file. The records are otherwise held in a random manner, and so there is no obvious relationship between contiguous records. Random access mode provides the user with the capability of obtaining access to any record directly without having to search through the file. In its true meaning, random access can apply only to a random file, although randomly stored files can be accessed serially.

Random storage is normally employed for master files for which the transactions cannot be sorted. Each transaction obtains access directly to the file record with which it is associated by its key being used to create the address by means of an address generation algorithm.

The advantages of random processing are:
1    The transactions do not require to be sorted, so saving time.
2    No indexes are needed, thus saving storage space.
3    An address generation algorithm is a faster method of finding a record than are most other methods.
    The disadvantages are:
1    If the file activity is high, random access is more time-consuming than sequential access.
2    Considerable wastage of storage may occur owing to some storage areas overflowing while others are only partially filled (Section 7.4).
3    A change in the values of the set of keys tends to aggravate the above situation. This may occur with a volatile file to the extent of demanding frequent housekeeping runs.
4    The records cannot be output sequentially except by sorting them or inputting a list of the keys in the required sequence.

## 6.5    Comparison of data storage media

When discussing storage devices, serial access is analogous to a serially stored file in that the data is accessible only by moving along the medium searching for the required record. Serial storage media is typified by and largely consists of magnetic tape. Magnetic tape cannot be addressed, i.e. the computer is unable to go directly to a piece of stored data because there is no means of locating it. In order to gain access to a particular record, it is necessary to read a block of records from the magnetic tape into main store and then search through the block for the wanted record.

A direct access storage device, such as a magnetic disk unit or a magnetic drum, is far more accessible. The computer is capable of reading a piece of data from a selected area of the storage medium into main store.

Thus, as indicated in Figure 6.1, a direct access device can be used more flexibly as regards the modes of storage and access of its stored files.

A considerable number of data storage devices, covering a wide range of speeds and capacities, is currently available. It is not the intention here to make a fine comparison of the various models, since full technical information about them is readily available from the manufacturers. During the final stages of systems design, the analyst becomes closely involved in making comparisons between competitive equipment, and he will then find it necessary to obtain the latest hardware specifications from the suppliers.

Before this stage however, the new system has to be designed, at least in broad principle, so as to make it possible to choose from one or more of the following categories of storage device:

1    Serial access devices
     – magnetic tape.
2    Direct access devices with removable storage units
     – exchangeable disk cartridges.
3    Direct access devices with non-removable (fixed) storage units
     – fixed disk units
     – magnetic drums.

These devices have a multitude of characteristics and to compare them all is beyond the scope of this book. Their main characteristics are however worth comparing in order to demonstrate their suitability for a given system.

*On-line data volume*    This is the volume of data that is accessible to the computer's processor at any one time without human intervention, i.e. on-line.

The devices in categories 1 and 2 above have no limit to the total volume of data that can be stored on the storage medium itself. There is, however, a limit to the volume that can be on-line at any one time; this limit is imposed by the characteristics of the device and by the number of units of the device fitted to the computer.

Typical on-line data volumes are as follows:

*Magnetic tape.* Up to 100 million characters, depending on the packing density of the tape and the length of tape per reel. More data can be made available on-line by using several tape decks simultaneously, but this is not always a practical proposition.

*Removable/disks* Up to 300 million characters per disk unit.

*Fixed disks* Up to 800 million characters.

*Magnetic drums* These may have relatively small capacities; the usual range is between 100,000 and 5 million characters.

*Access time* The term 'access time' as applied to magnetic tape is not really meaningful. This is because magnetic tape operates in an inherently serial manner and therefore any one record is not accessible in a time that is relevant.

Access time is the time taken for the computer to gain access to the data stored on the device, and consists for the most part of the time occupied by mechanical movements of one kind or another.

*Removable disks* These are continuously rotating discs, upon the surfaces of which the data is stored; access is obtained by read/write heads mounted on either fixed or movable arms. The access time is made up of head positioning time and the disk, rotational delay time, and lies between 25 and 70 milliseconds.

*Fixed disks* The principle of operation is similar to removable disks; their larger size tends to reduce their speeds of operation, giving access times down to 30 milliseconds.

*Magnetic drums* These are smaller and faster than the above devices, and there is no movement of the read/write heads. The access time is due entirely to the drum's rotation and varies from 0 to 20 milliseconds.

*Transfer rate* This is the rate at which data can be transferred from the storage device into core store, or vice versa. It is of particular importance in relation to the employment of the device for the sorting of records. The approximate transfer rates are below:

Magnetic tape – 30,000 to 300,000 characters per second.

Removable disks – 30,000 to over 1 million characters per second.

Fixed disks – up to 800,000 characters per second.

Magnetic drums – up to over 1 million characters per second.

## 6.6 Modes of processing

In the present context processing means operating on a file in order to make changes to it. The resultant set of records in the file differs from the initial set. The changes may be either in regard to the data items within the records or as regards the set of records itself. These are known as file updating and file amendment respectively.

There are two main ways in which file records are processed — reconstruction mode and overlay mode.

## Reconstruction mode

This method entails reading the file records from backing storage into main store and, after updating or amendment, writing them to another area of backing storage to form a 'reconstructed' file. Thus the original file is preserved in its original locations and two files then exist. These are known as opening and closing files or brought-forward and carried-forward files.

Reconstruction mode is the only processing mode applicable to magnetic tape. With this storage medium there is no choice but to read records from one reel and write to another. It is also possible to use reconstruction mode with direct access devices, such as magnetic disks. The main reason for doing so is in order to obtain a high degree of security. This is achieved because after the run the original records still exist as they stood, thus facilitating a re-run in case of trouble having occurred. Trouble takes the form of damage to records, errors in the input data, overflow of data areas (Section 7.4), and the accidental overwriting of the brought-forward records before they have been properly updated.

It is necessary also to reconstruct the unchanged records in order that the file is complete at the end of the run. This can occur automatically with sequential processing but with random processing, the unchanged records have to be transferred as they stand at the end of the run.

With sequential processing, the master record is extracted from its original location, updated in main store by one or more transactions, and then written to the appropriate location in the reconstruction area. With random processing, the transactions appertaining to a given master record are dispersed, and so the original record is transferred after its first updating and thereafter updated in the reconstruction area.

Reconstruction mode has the disadvantages of requiring double the storage space for the file, i.e. the original area and the reconstructed area. It has the advantage that expanded records and an increased number of records are catered for fairly easily. With sequential or indexed-sequential storage, the updated or amended records are written in turn to the reconstruction area. This means that they merely take up the next space available. Random storage implies that a fixed amount of space is allocated to a group of records and expansion beyond this means that special measures must be adopted (see overflow in Section 7.4).

*Overlay mode*

With overlay mode updated records are written back to the original location in the storage area. Thus the original data is lost as soon as an update occurs and consequently a security hazard exists unless a copy is dumped beforehand.

Overlay mode is not possible with magnetic tape nor is it practicable on a direct access device with variable length records or an expanding file. The original record might be overlayed many times during an updating run and so, as stated earlier, it is wise to dump a copy of the original file on to another area of disk or to magnetic tape at the start of the updating run. The overlaying risk is also reduced by a method known as 'after-state copying' (Section 7.5).

Advantages of overlay mode are that it is possible to update a small proportion of the records without disturbing the remainder, and a lesser amount of storage space is required.

## 6.7 Data bases

A data base consists of sets of interrelated data held in a computer's storage in such a way as to serve as many applications as possible in an optimum fashion. The organization and control of the various data is common to all sets and allows them to be substantially independent of the programs using them.

In the early years of computer usage for business, the computerised applications were largely independent of one another. Suitable files could therefore be created without fear of duplication or confliction of data. With increasing computerisation, the applications became more likely to require the same or similar data in their file records. This not only wasted storage space but also allowed the possibility of conflicting values of the same data items. These discrepancies could be caused by different routines being employed to update and amend the professedly identical data. In particular, the phasing of routines caused some data to be more up-to-date than other data.

Another difficulty occurred when several programs needed to locate and retrieve data from the same file records. At best, all these programs would use the same accessing method and so a common sub-routine could be used. Nevertheless if one program demanded an alteration to the contents of a set of records, all the other programs had to be modified to allow for this. At worst, the various programs had been written to use different accessing methods and an alteration to data structure then caused considerable work and difficulties in modifying the programs.

Thus it is obviously beneficial if the structuring and accessing of data can be standardised. The more common accessing methods are described in Chapter 7.

*Characteristics of a database*

1   *Data/program/medium independence*   This means that the data structure and application programs using a database can each be changed without affecting the other, i.e. are logically independent, and that the programs are isolated from changes made to the storage medium holding the data, i.e. physical independence.

Thus if it is decided to change from a sequential mode to a random mode of storage, the application programmer need not know. He merely calls for a piece of data and the database management system (DBMS) delivers it to his program. Similarly, a changeover from, say, a disk system to a drum system does not concern him. The database management system (DBMS) is capable of finding and transferring data wherever it is stored and on whatever type of medium.

2   *Flexibility of data*   The applications programmer is concerned with logical files, in other words in manipulating file records in the way he imagines them to exist. Individual programmers may each have different ideas as to the contents and layout of a particular logical file, each one seeing it according to his own needs. The database is structured so as to allow for this, and thus be capable of presenting a program with the data items it requires. These are made to appear by the DBMS as if constituting the record(s) that the application program calls for. Another program may want a different set of data items, some of which are the same as those wanted by the first program.

It is obviously advantageous if several different logical files can be created from the same stored data without the programmer having to organise this and without the data being replicated.

3   *Minimal redundancy*   This means that, as far as practicable, there should be only one copy of each piece of data stored in the database, i.e. no unintentional replication of data. This may be a far cry from some existing sets of files, especially those on magnetic tape, for which it is necessary to replicate data items so that they are always on-line to the computer when needed.

By having only one copy the possibility of conflicting values of data items is eliminated, and storage space is reduced. Where it is disadvantageous to eliminate duplications, the DBMS must ensure that no inconsistencies are apparent.

4   *Efficiency*   The database must be organised so that it can be

searched for a variety of data in response to different demands. The same data may be utilised both by a batch processing system and by a real-time system, each demanding different information based upon that data.

One of the main problems is in maintaining a low response time, especially if a man-computer dialogue is employed [4.13]. Maintaining efficiency often necessitates moving the more popular data into the more accessible storage areas. This has to be done dynamically and automatically by the DBMS. The DBMS has therefore to monitor the demands for data continuously and reorganise it accordingly. This process is sometimes called data migration.

Another aspect of efficiency is that a database should be tunable. This means that it is capable of being adapted in order to handle most efficiently the type of processing demanded more frequently.

5   *Security (integrity)*   Data has to be safeguarded against accidental or intentional destruction, damage or loss. These hazards may occur as a result of hardware failure, software errors, operating errors or interference. The database should hold the data in such a way that loss or damage can be repaired.

Security also includes guarding against fraud and the misuse of data. This involves checking on the identity of terminal users and their authorisation to gain access to certain data. The database has to be auditable, tamper-proof and monitorable.

6   *Privacy*   If a data base holds data concerning individual persons, and perhaps organisations, the right of data privacy may exist. This is a more of a socio-legal problem than a computer one but nevertheless the database needs to cater for data privacy.

*Database management systems*

In order to achieve the characteristics described above, it is necessary to have some means of controlling the database. The database management system (DBMS), mentioned earlier, does this − or at least goes some way towards this end. The DBMS is a complete piece of software held permanently in the computer and frequently brought into action. The main characteristics of a DBMS are:
1   To be capable of interfacing with the application programs. This entails accepting a requirement as defined by a data name and key value, and finding and presenting the relevant data. The DBMS interprets a requirement by means of its schemes (program and logical data descriptions) into the physical storage of the data.

2　To be capable of interfacing with the computer's operating system.
3　To be able to translate the languages used to define the schemas, logical data, physical data and program languages.
These languages include:
*a*　the data manipulation language connecting the DBMS with the application programs using it;
*b*　the data (subschema) description language informing the DBMS of the logical data structures used by an application program;
*c*　the data (schema) description language informing the DBMS of the overall (global) logical data in use by all the application programs;
*d*　the physical data description language; this maps the global logical data on to the physical storage media in use.

## 6.8　References and further reading

6.1　MARTIN, *Computer database organization,* Prentice-Hall (1975).
6.2　'Aspects of database management', *Data Processing* (March-April 1971)
6.3　'Currents in database thinking', *Data Systems,* (Jan-Feb 1974).
6.4　'Towards content addressing in databases', *Computer Journal* (May 1972).
6.5　'Feature analysis of generalized database management systems', *Computer Bulletin* (April 1971).
6.6　'Good data management', *Data Systems* (Jan-Feb 1974).
6.7　'Self organizing data management system', *Computer Journal* (Aug 1974).
6.8　'Robot – an independent DBMS', *Data Processing* (Sept-Oct 1973).
6.9　'Database – the ideas behind the ideas', *Computer Journal* (Feb 1965).
6.10 'A model of self-organizing data management system', *ibid* (February 1974).
6.11 'The thinking behind a database management system', *Data Processing* (September - October 1974).

# Chapter Seven
## *Direct access storage*

### 7.1 Storage concepts of direct access

The storage concepts associated with direct access storage devices are of two kinds — hardware concepts, and software or logical concepts. The former are those innately connected with the physical electro-mechanical design of the device. Logical concepts concern the way in which the storage medium is used to hold data and facilitate access to it. From the systems and programming viewpoints it is the logical concepts that are the more important, but before describing these in detail, a further brief explanation of the hardware concept is worthwhile.

*Hardware concepts of disk devices*

A magnetic disk unit consists of a stack of physically connected rotating disks, on the surfaces of which the data is recorded in the form of magnetized spots. Each recording surface consists of a number of concentric bands (tracks) divided into blocks, as shown in Figure 7.1. A block is the smallest amount of data that can be transferred to or from the device at a time; in some devices the band must be transferred as a whole, and is therefore equivalent to a block in this respect. It is usual to transfer one or more blocks at a time into main store, the precise number depending upon the logical layout of storage and the amount of main store available to receive them. Each band on a surface holds identification data in addition to the data for processing.

The read/write heads are all physically attached to one arm and move synchronously across the disk surface. These are known as movable heads. Some models of disk units have a number of fixed read/write heads permanently positioned over certain bands (one head per band).

Normally the stack of disks (known as a pack, module or cartridge) is removable from the drive (spindle and control unit). This means that one lot of data can be replaced by another lot held on another cartridge. Some models of disk units also have a number of fixed (non-removable) disks mounted on the same spindle as the removable cartridge. This arrangement enables permanently required data, such as the computer's operating system, to be readily available.

If necessary, several drives are attached to one computer, either each with its own control unit (channel) or sharing the one control unit.

## Hardware concepts of drum devices

A magnetic drum comprises a rotating cylinder around the circumference of which are a number of tracks holding data in magnetic form. There is a read/write head for each track and data is transferred as it passes the head. Magnetic drums are not removable and, as seen from Section 6.5, have lower storage capacities but higher transfer rates than magnetic disks.

## Logical concepts of direct access storage devices

The systems analyst need not concern himself too deeply with the hardware of direct access storage devices, but should think in terms of three software concepts. These are the cylinder (seek area), the bucket and the record; records are described in Section 6.2 and appear again in this chapter.

*Cylinders (seek areas)*   Referring to Figure 7.1, it can be seen that the bands on the surfaces of the disks comprise a number of imaginary cylinders. The data held in the bands that form the 'surface' of each cylinder can be accessed by the set of read/write heads by positioning them once only. In other words, each band is accessed by one of the heads, the computer then switches electronically to another head and another band is accessed. This process continues through all the bands until the whole cylinder has been accessed without further head movement.

This concept of cylinders is important because movement of the read/write heads represents a significant portion of the access time,

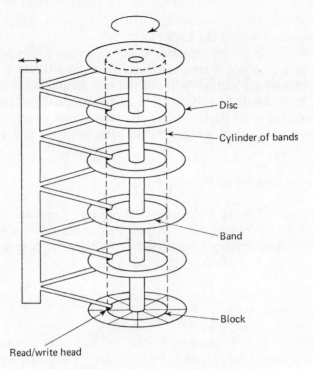

Labels on figure:
- Disc
- Cylinder of bands
- Band
- Block
- Read/write head

**Figure 7.1** *Cylinder concept of disk units*

and should therefore be minimised.

The cylinder concept as applied to a magnetic drum is the complete set of tracks because the data thereon is accessible without any time delay caused by head movement.

*Buckets*   A bucket is a logical portion of data that is transferred in its entirety to or from the disk. It can consist of one or more blocks, but is normally of constant size within a file. The physical layout of a bucket need not be considered; it is satisfactory to think of it merely as a logical area of storage. During processing, a bucket at a time is transferred into main store, and it is the amount of main store available to accommodate the bucket that largely determines the bucket size. The bucket of a drum comprises one or perhaps a few tracks.

Thus, from the systems design aspect, disk storage and drum storage can be regarded as made up of cylinders within which are buckets. Typically, a disk cartridge consists of 200 cylinders of 40 buckets, each bucket holding 9000 bytes. Each bucket is loaded with logical records, the number of which is dependent on their size. In order to access a record, it is necessary to determine which cylinder it is in and the bucket within that cylinder. Thereafter the bucket is searched serially or, occasionally, it is possible to also deduce the record's position within the bucket.

In the succeeding sections some of the more common addressing and searching techniques are described briefly. If a DBMS is employed, these techniques are utilised without the system analyst or programmer being aware of this. In cases where no DBMS is used, these techniques have to be designed into the system and so it is important that the systems analyst is cognisant of their purposes and advantages.


## 7.2   Record addressing techniques

Locating a record by an addressing technique presupposes that the record's address is calculated or deduced from the value of its key. Thereafter the computer is instructed to transfer the data in that location's address to wherever it is required.


*Self (direct) addressing*

This method is suitable for determining the addresses of fixed-length records in a sequential file, and in which the keys form a complete or almost complete range of consecutive numbers. Under these circumstances, it is possible to find the address of a record by a simple modification to its key.

Suppose, for example, the address is required of the record with key 43746 in a file of 16,000 records, with keys 30001 to 46000. The file is held in 20 cylinders each of 100 buckets. The buckets are, in effect, addressed as numbers 2001 to 4000, and each bucket holds 8 fixed-length records.

The steps are:

1    Divide the wanted record's key minus the lowest key by the number of records per bucket:
$$(43746 - 30001) \div 8 = 1718 \text{ and } 1 \text{ remainder,}$$

2    Add the first bucket number to the quotient to give the wanted record's bucket; the remainder plus one is the record's position within that bucket:
$$2001 + 1718 = 3719$$

Thus the wanted record is the second in bucket number 3719.

This method is not very realistic in that large files with a complete range of consecutive keys are few and far between. It is more likely to be useful for smaller files as might be stored on magnetic drums. Where a file has a limited range of non-volatile items, it may be worth re-numbering them in order to employ this method.

Self-addressing has the following advantages:

1    The key leads directly to the record, and so need not be stored

within the record. If the key is required for other purposes, it
can be easily derived from the address of the record.
2    No index tables are necessary, thereby saving storage space and
the time to search them.
Self-addressing has the following disadvantages:
1    The records must be of fixed length, otherwise a trailer record
chained to the main record is needed.
2    Non-existent records, whose keys would lie within the range
covered by the file, must have storage space allocated to them.
This is wasteful of storage, but this may be acceptable if there
is plenty available.

*Partial addressing*

This is somewhat similar to the previous method except that only the
bucket number is derived from the key. The wanted record is then
found by searching through the records in the bucket. This method is
used with a sequential file that has a fairly even spread of keys; the
file can contain fixed- or variable-length records, and is preferably of
low volatility.
    The bucket number of a wanted record is found as follows:
1    Calculate average bucket spread = largest key in file minus
smallest key plus one, divided by number of buckets.
2    Subtract smallest key in range from wanted key.
3    Divide result of 2 by average bucket spread, rounding to nearest
integer.
4    Add lowest bucket number to result of 3 to give bucket number
of wanted record.
5    Search this bucket for the wanted record, moving forward
through successive buckets if the stored records' keys are
less than the wanted key, and vice versa.

**Example**    The bucket number of the record with key 3157 is required from a file of 500 records with keys in the range 2501 to
3500, stored in buckets numbered 550 to 649, i.e. 100 buckets.
1    Average bucket spread = (3500 − 2501)/100 = 10
2    3157 − 2501 = 656
3    656 ÷ 10 = 65
4    550 + 65 = 615 = required bucket number for start of search.

*Address generation (randomizing)*

Address generation is used with random files and uses an algorithm
(set of arithmetic operations) in order to form the address (bucket

number) of a record from its key. The address generation algorithm involves a few simple arithmetic operations on the key, and the nucleus of the problem is to find an algorithm that creates a uniformly spread range of bucket numbers from a given range of keys. Ideally the same number of records are assigned to each bucket. Unfortunately no algorithm can achieve this perfect situation, with the result that some buckets overflow while others are partly empty. The degree of common bucket numbers (synonyms) generated depends upon a number of factors:

1   The bucket capacity, i.e. the average number of fixed or variable length records that can be held in a bucket.
2   The file packing density, i.e. the proportion of the file's storage area that is occupied by records.
3   The structure of the key set, and the algorithm employed to operate upon it.

The main address generation algorithms are described below; it should be borne in mind however that these may be combined together in order to provide the best arrangements under given conditions.


*Prime number division*

By dividing a key by the number of buckets available, a remainder is formed lying between nought and one less than the number of buckets This remainder, when added to the first bucket number, could be used as the bucket number of the record. However, owing to the bias that tends to be present in a range of keys, the remainders so formed are by no means evenly distributed. This is especially so if the keys have a tendency towards only certain values of their final digits, which is not unusual.

If it is known that the bias applies only to the final digit of a key; this can be effectively eliminated by exchanging the positions of the digits before dividing. In practice, however, this is likely to merely bring another biased position into the final position.

A better solution is to use a prime number as the divisor. This should be the highest prime number below the number of available buckets for the file. The efficacy of a particular prime number should be checked by using a simple program to analyse the bucket assignments, moving to lower prime numbers if a more uniform spread is so obtained.

In order to determine the extent of bias within a set of keys, it is possible to employ a 'digit analysis program' on a computer. This makes a count of each digit value in each position of the key; these counts can then be inspected in order to decide which of the algorithms is the most suitable and which digit positions are best used in the methods described below.

*Extraction or truncation*

In this method the most random digits of the key are extracted and formed into a bucket number. If these happen to lie together at one end of the key, the method is called 'truncation'.

**Example**  Key 395628, using first and alternate digit positions would give bucket number 352; by truncating the last three, the bucket number would be 628. The first bucket number is added as with all these methods.

*Folding*

The key is split into two or more parts; these are then added together in order to obtain a higher degree of randomness.
**Example**  Key 265304,
1    Split into two parts, 265 + 304 = 569
2    Split into three parts, 26 + 53 + 4 = 83.
3    Using alternate positions, 250 + 634 = 884.
   Another version of folding pivots the key around its mid-point and adds the two lots of digits, some of which have thus reversed in order.

**Example**  Key 379584:
1    Pivoting the last three digits 379 + 485 = 864.
2    Pivoting the first three digits 973 + 584 = 557 (ignoring the most. significant digit of the sum).

*Squaring*

The key is squared and a portion of the result used as the bucket number. This is multiplied by a factor to bring it to the number of buckets available.

**Example**  Key 438 with a range of 70 buckets.
$438^2$ = 191844, taking 4th and 5th digits gives a bucket number of 84 × 0.7 = 58 (rounded down).

*Combinations of extraction, folding and squaring*   The above methods may be combined in any way that increases the randomness of the result. The result still tends to be suited to cases where the number of buckets is a multiple of a power of ten, but this can be overcome by using a multiplying factor as in the last example.
   *Alphabetic keys*—the address generation algorithms described in the preceding pages have all been concerned with numeric keys, but

this does not mean that alphabetic keys cannot be used to form bucket numbers. The general principle is to convert the alphabetic characters in a key into numeric equivalents before injecting them into an algorithm.

The most straightforward method is to replace the letters A to Z by the digits 01 to 26 respectively. Alternatively the last digit only is used but this loses some of the key's randomness.

## 7.3 Record locating techniques

*Partial indexing*

This method of locating records is the most commonly employed because it is suitable for almost all types of key ranges encountered in practice. It makes use of an index in which are held the highest keys of the records in each bucket, and although strictly speaking the records need not be in sequence within each bucket, it is usually more convenient to have them so for practical purposes. Partial indexing corresponds to selective-sequential and indexed sequential storage. The index described above is known as a bucket index, and it is convenient to hold it in the first bucket of a cylinder, transferring it into core store before use. It may be possible to hold the index permanently in core store if there is sufficient space, such as there well might be if only one cylinder exists. If a number of cylinders are in use for one file, a cylinder index is held permanently in core store. This index contains the keys of the highest records within each cylinder, and thereby enables the appropriate cylinder to be found. The bucket index from this cylinder is then used to find the wanted record's bucket, and this is searched sequentially to find the wanted record.

It will be realized that if the file is being processed against movements that are in the same sequence as the file, the cylinder index is not really needed because each cylinder is used in turn anyway. The same argument also applies to bucket indexes in a file with high activity.

*The advantages* of partial indexing are as follows:
1    It is straightforward to apply — manufacturer's software is available for creating and searching the indexes.
2    Wasted storage is minimal, up to 100 per cent packing density being possible.
3    Variable-length records can be used.
*The disadvantages* are:
1    Index tables occupy space in store.
2    Time is taken to search the indexes.

## Binary searching (binary chopping)

Whereas with partial indexing there is a limited index containing one entry per bucket, binary searching is normally used with a large index containing one entry per record. This entry holds the key of each record, so that having found the wanted record's key in the index, it is possible to go directly to the record itself in disk storage.

The principle behind binary searching is to compare the wanted record's key with the key at the mid-point of the index. Depending whether the wanted key is greater or lesser than the mid-point key, the next comparison is made with the index key situated at the three-quarter or quarter point of the index. This principle is continued until identity is found between the index key and the wanted key. This means that each comparison halves the area of index left for searching and consequently the number of comparisons has a logarithmic relationship to the size of the index. If the number of entries in the index is $E$, the maximum number of comparisons is $\log_2 E$, and the average is one less than this. A simple example of binary searching is shown in Figure 7.2.

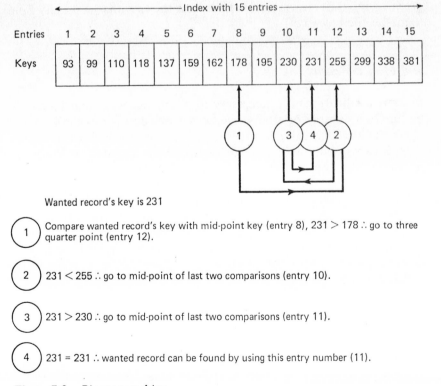

Wanted record's key is 231

1. Compare wanted record's key with mid-point key (entry 8), $231 > 178 \therefore$ go to three quarter point (entry 12).

2. $231 < 255 \therefore$ go to mid-point of last two comparisons (entry 10).

3. $231 > 230 \therefore$ go to mid-point of last two comparisons (entry 11).

4. $231 = 231 \therefore$ wanted record can be found by using this entry number (11).

**Figure 7.2** *Binary searching*

It will be noted that in this example, the number of entries in the index is such that it is always possible to move to a mid-point in the remaining keys; this is because the total number of entries is a power of 2 minus 1. In practice this fortunate state does not usually pertain; it is however possible to deal with other numbers of entries by using one of the following alternative methods.

1    Make the number of entries into a suitable amount by inserting dummy entries at the beginning or end of the index. These must contain dummy keys the values of which are less than the first real key or greater than the last real key respectively.

2    Modify the binary searching sub-routine so that it ignores 'entries' that are outside the limits of the index. Before each comparison the location is tested; if it is greater than the highest location of the index, the subroutine does not extract the key but proceeds as if it was greater than the wanted key. If the location is less than the lowest location of the index, the opposite logic applies.

Binary searching is not normally employed with sequential transactions unless the file activity is low; it is more suitable for sporadic access to a sequential file. When using a disk device, the index generally resides in core store, but with a magnetic drum the searching is carried out among the records themselves. This is practical with a drum owing to its lower access time, and the fact that it usually holds a smaller file anyway.

*Multi-key accessing*

In some applications it is necessary to obtain access to a given record by one of a number of alternative keys. Consider, for example, a stock file in which there is one record per item stocked. Each of these records could hold three different unique numbers that might be used as the key for locating it, i.e. part number, bin number, drawing number. The stock file is held on a direct access storage device sequentially by part number because this suits the majority of the processing runs. However, it is necessary in some runs to access the records by using the bin numbers, and in others by using the drawing numbers. This cannot be done simply by attempting to create the required bucket number from either the bin number or the drawing number, because the file layout is controlled by the part number only.

The problem is solved by using address generation algorithms for the bin number and the drawing number that direct the computer to an index location instead of the actual record. Thus the bin number directs the computer to one location of the index, and the drawing number to another (see Figure 7.3). Both these locations contain the bucket number of the record to which the keys appertain. As it is

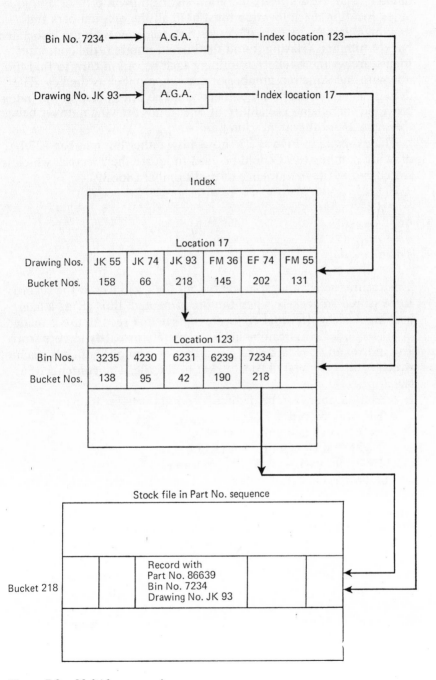

**Figure 7.3** *Multi-key accessing*

unlikely that the address generation algorithm will provide a unique index location, each location must hold all the bin numbers and drawing numbers that are directed to it, together with the associated bucket numbers. Having found the bucket number, the computer then searches the bucket, examining each record in turn to find the one with the same bin number or drawing number as the key. The drawing numbers and bin numbers could be intermixed in the index provided there is no possibility of one item's drawing number being the same as another item's bin number.

The example in Figure 7.3 shows how either bin number 7234 or drawing number JK93 could be used to locate their record, which is stored in the file in sequence of part number (86639).

*Tree searching*

This technique is applicable to random files, and causes the records to be stored adjacently when the file is created, thus giving a high packing density. In order to locate a particular record, use is made of a 'tree index' that can be held completely or partly in core store. The index can be regarded as consisting of a number of cells, each of which is directly related to a bucket in the file. The contents of a cell are:

Cell number (can be deduced from its core location).
Key of a record.
Left branch cell number (LBC).
Right branch cell number (RBC).
Deletion marker (DM) (if applicable).

The principle of tree searching is to compare the wanted record's key with an index key (starting with the first in the index), and then to go to either the left branch cell or the right branch cell depending on the result of the comparison. This procedure is continued until a comparison gives identity between the wanted record's key and the cell's key. The bucket number is obtained by dividing the cell number by the number of records per bucket — which must be fixed — or alternatively holding the bucket number in the cell.

By referring to Figure 7.4, the procedure for locating a key can be seen. The list of keys shows the order in which they have been put into the file; this order is random according to the key and in sequence of cell number because the cells are taken in turn. As each key arrives it is put into the first vacant cell, and its cell number inserted as the left or right branch cell number of the cell from which it branches. The logic for locating (or inserting) a key is as below. This example explains the steps in searching for key 640 in Figure 7.4.

TREE STRUCTURE

TREE INDEX

| Cell no. | Key in cell | Left branch cell no. | Right branch cell no. | Deletion marker = 1 |
|---|---|---|---|---|
| 1 | 547 | 3 | 2 | |
| 2 | 603 | 8 | 6 | |
| 3 | 175 | | 4 | |
| 4 | 258 | 14 | 5 | |
| 5 | 369 | | | |
| 6 | 691(662) | 11 | 7 | (1) |
| 7 | 752 | | 9 | |
| 8 | 588 | | | |
| 9 | 789 | 10 | | |
| 10 | 766 | (16) | | |
| 11 | 640 | 12 | 13 | 1 |
| 12 | 626 | | | |
| 13 | 649 | | | |
| 14 | 212 | (15) | | |
| 15 | (195) | | | |
| 16 | (760) | | | |

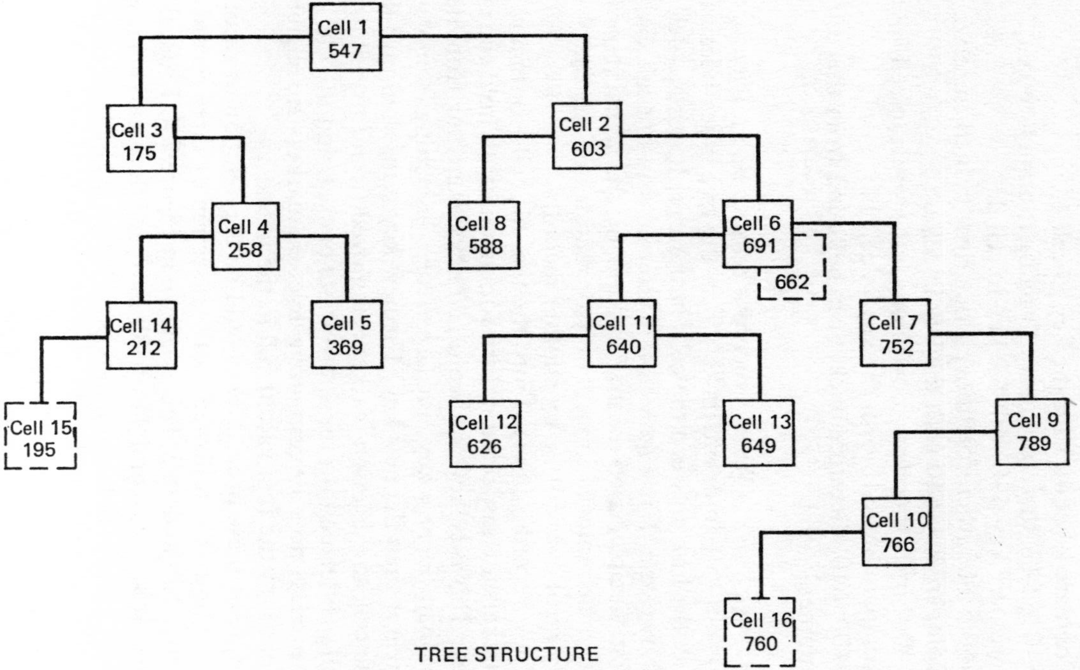

**Figure 7.4** *Tree searching*

*Step 1*   Compare key 640 with the key in cell 1,
           640 > 547, therefore move along right branch to cell 2.
*Step 2*   Compare key 640 with the key in cell 2,
           640 > 603, therefore move along right branch to cell 6.
*Step 3*   Compare key 640 with the key in cell 6,
           640 < 691, therefore move along left branch to cell 11.
*Step 4*   Compare key 640 with the key in cell 11,
           640 = 640, therefore bucket can be found from this cell
           number.

*Amending a tree index*   When a record is removed permanently from
the file, it is not safe to delete its key from the tree index because this
would break the link between the cells in the tree. Instead a 'deletion
marker' (shown as a 'I' in Figure 7.4) is inserted into the cell so as to
indicate that the associated record is redundant; the actual record itself can either be removed immediately or retained on the file
temporarily until overwritten by another record.

When a new record is inserted into the file, the cells containing
deletion markers are inspected in turn for eligibility to hold the new
record's key. This is achieved when (*a*) *all* keys branching from its
left are less than the new key, and (*b*) *all* keys branching from its
right are greater than the new key. The new key is inserted into the
first eligible cell, the deletion marker is removed from this cell, the
record itself is inserted into the appropriate bucket, and the old
record removed. If no cells containing deletion markers are eligible
the key is put into the first vacant cell in the index.

An example of these procedures would be the deletion of records
with keys 691 and 789, followed by the insertion of three records
with keys 662, 195 and 760. This is shown in Figure 7.4. From this
it can be seen that cells 6 and 9 initially have deletion markers inserted (shown as 'ones'), indicating that their keys are eligible for replacement. New key 662 fulfills the requirements for replacing key
691, and the deletion marker is removed (shown bracketed). New
keys 195 and 760 are unsuitable as replacements and so are inserted
at the end of the index.

## Advantages of tree searching
1   High file packing density is possible even with volatile files
    because deleted records are replaced by new ones.
2   Fairly rapid access to a random file.
3   Records may be fixed or variable length, but if variable, allow-
    ance must be made for overflow.

## Disadvantages of tree searching
1   A large index is necessary, and arrangements must be made for
    transferring a section of it at a time into core store.
2   Time is taken in searching the index.

## 7.4 Distribution of records

In the previous sections of this chapter a number of methods for locating records have been described. These have resulted in the determination of the number of the bucket in which the wanted record should normally reside. This bucket is known more accurately as the 'home bucket' of the record, and is the one to which the record is initially assigned. If, for the reasons that are described below, a record cannot be accommodated in its home bucket, it is known as an 'overflow record' and is re-assigned to an 'overflow bucket'.

*Sequential file overflow*

When sequential files are created, records are assigned sequentially to buckets and, at the same time, an index is usually created. No matter whether the records are of fixed or variable length, they are always accommodated at this stage within their home buckets. Subsequently, during file updating and amendment runs, records are expanded and inserted, with the result that they may overflow from their home buckets.

There are two main causes of bucket overflow in a sequential file:

1   The insertion of records into buckets that are almost full. This condition can apply to fixed or variable length records, except when self-addressing is used with fixed length records.
2   The expansion of variable length records. These sometimes have more items inserted during an updating run with the result that the home bucket can no longer accommodate the expanded record.

*Random file overflow*

This occurs for another reason in addition to those mentioned above. It will be remembered that with random files the home bucket numbers are created by an address generation algorithm. This method always produces synonyms, and any mal-distribution of these between home buckets also causes overflow to occur. Thus with random files, overflow buckets are needed even during the file creation stage.

*Overflow areas*

Since bucket overflow is inevitable, some space on a direct access storage device must be allocated for overflow records. The overflow areas are separate from the home areas and may be either *(a)* within the same cylinders, or *(b)* all concentrated into one cylinder. These are known as 'cylinder overflow areas' and 'independent overflow areas' respectively. An advantage of having an independent outflow area is that less space need be reserved for overflow. A disadvantage is that more access time is taken in getting to overflow records since these are not in the same cylinder as the one immediately available. A suggested approach is to have cylinder overflow areas that are large enough to contain the average amount of overflow during updating, and to have an independent overflow area for use when a cylinder overflow area is full.

## Progressive overflow

This is the most straightforward but least efficient technique for dealing with overflow records, and is used only with random files. When an overflow occurs, the overflow record is inserted into the next higher bucket with sufficient space for it. The link between the home bucket and the overflow bucket is simply its nearness. When searching for a record, the home bucket is examined first and if this does not hold it, successive buckets are examined until the wanted record is found.

## Chaining

This method is so called because each home bucket contains a 'chaining record'; this is simply a small record at the beginning of each bucket that holds the number of the associated overflow bucket.

*Overflow area chaining*  The chaining record of each home bucket directs the search to the first bucket in the overflow area that could hold the wanted record. If this overflow bucket does not in fact hold the record, the search continues from this point onwards. This method is illustrated in Figure 7.5.

An alternative is to also insert a chaining record into the overflow bucket if this becomes full. The search then continues via the chaining record until the wanted record is found.

With the above chaining methods, expansion of a record beyond the capacity of its bucket (home or overflow), simply entails moving it into the next overflow bucket that can accommodate it. No alteration of the chaining record is necessary, making the procedure straightforward, but chaining can lead to inefficient utilization of storage unless care is taken to re-fill the home buckets with new records, or the file is re-organized by regular housekeeping runs.

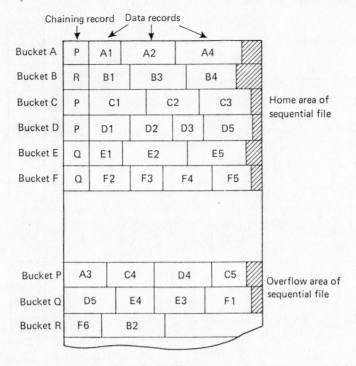

Chaining record · Data records

| | | | | | |
|---|---|---|---|---|---|
| Bucket A | P | A1 | A2 | A4 | |
| Bucket B | R | B1 | B3 | B4 | |
| Bucket C | P | C1 | C2 | C3 | |
| Bucket D | P | D1 | D2 | D3 | D5 |
| Bucket E | Q | E1 | E2 | E5 | |
| Bucket F | Q | F2 | F3 | F4 | F5 |

Home area of sequential file

| | | | | |
|---|---|---|---|---|
| Bucket P | A3 | C4 | D4 | C5 |
| Bucket Q | D5 | E4 | E3 | F1 |
| Bucket R | F6 | B2 | | |

Overflow area of sequential file

When the file was originally created, records were situated in their home buckets (as indicated by letter in key). During the course of several amendment/updating runs, insertions and expansions of records have resulted in overflow occuring and the re assignment of records as shown. The order in which the records were re-assigned was A3, C4, D4, C5, D5, E4, E3, F1, F6, B2.

**Figure 7.5** *Overflow chaining*

## Tagging

This technique is applicable to both sequential and random files, and is more suitable for use with variable length records. As shown in Figure 7.6, the principle involved is the replacement of a record that overflows by a 'tag' in its home bucket; the tag holds the record's key and the overflow bucket number. Thus, when searching for a record, if a tag is found in its place in the home bucket, the search goes straight to the overflow bucket indicated in the tag.

If subsequently an overflow record expands beyond the capacity of the overflow bucket, it is transferred to the next overflow bucket with available space. The tag in the home bucket is altered to hold the new overflow bucket number; no tag is necessary in the original overflow bucket. The gaps left in buckets by this transfer procedure are filled later by new records, or by a housekeeping run that not only closes up the file, but in the case of sequential files, completely re-distributes the file so as to eliminate tags.

111

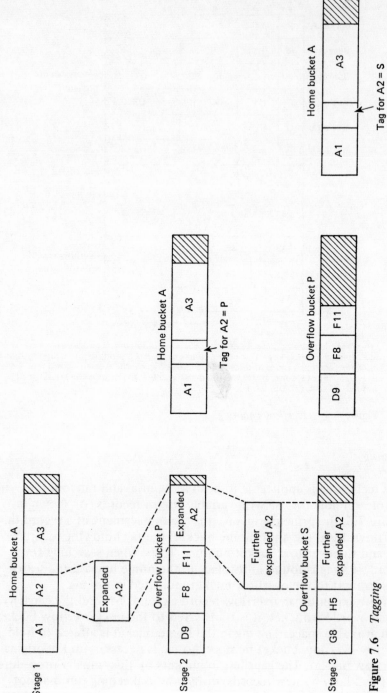

**Figure 7.6** *Tagging*

112

The procedure followed when inserting a record, for which the home bucket has insufficient capacity even for its tag, is to transfer another home record to an overflow bucket, and then to insert two tags into the home bucket.

The aforementioned overflow techniques are built into database management systems and also are available as software from computer manufacturers and software houses.

*Estimating the amount of overflow*

During the updating and amendment of files, new records are inserted, obsolete records deleted and current records expanded or contracted. In general with a fairly static file these changes tend to balance out over a period of time; thus, provided the file is re-organized regularly, there is no need to make a significant allowance for overflow. This is obviously not so with an expanding file, and it is useful to be able to approximate the amount of overflow storage that is needed when the file expands by a known amount. This can be found from Figure 7.7 in which is a table showing the percentage of inserted records that are assigned to overflow buckets. In order to use the table, it is necessary to find the average spare bucket space, ASBS (in records). If

$R$ = total number of records in the file,
$B$ = number of buckets in the file,
$r$ = average record size in characters,
$b$ = bucket size in characters,

then the approximate average spare bucket space is given by

$$\text{ASBS} = [b/r] - (R/B)$$

the square brackets indicating that this quotient is rounded down.

**Example** Suppose we insert another 24,000 records into a file of 100,000 records of average length 200 characters, stored in 20,000 buckets of 1500 characters. Then

$$\text{ASBS} = [1,500/200] - (100,000/20,000) = 7 - 5 = 2 \text{ records}$$

When 24,000 records are inserted,

$$\frac{\text{Number of records inserted}}{\text{Number of buckets in file}} = \frac{24,000}{20,000} = 1.2$$

Using the table in Figure 7.7 we find that 13.6 per cent, i.e. 3264, records can be expected to overflow. This is an interesting discovery because, at first sight, one might assume that as there is plenty of spare storage space, no overflow would occur.

Total storage space = 20,000 buckets |X 1500 characters   = 30 M characters
Over-all file size initially = 100,000 records X 200 characters = 20 M characters
Spare storage space   = 10 M characters
Insertions = 24,000 records X 200 characters   = 4.8 M characters

Thus in spite of the fact that only 48 per cent of the spare storage space is actually required, overflow occurs nevertheless.

*Random file distribution*

It will be recalled from Section 7.4 that an address generation algorithm is an attempt to operate on a set of keys in order to obtain an even spread of bucket numbers. The degree of success achieved depends upon both the algorithm and the structure of the set of keys. With a random file, the algorithm is used not only to retrieve records

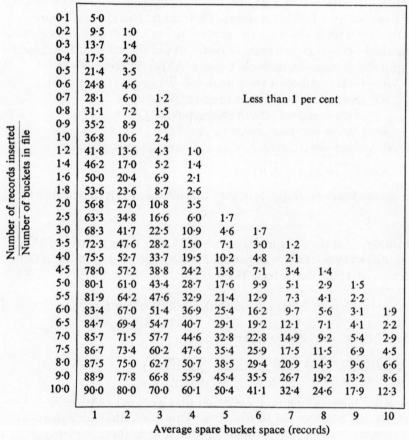

$\dfrac{\text{Number of records inserted}}{\text{Number of buckets in file}}$

| | 1 | 2 | 3 | 4 | 5 | 6 | 7 | 8 | 9 | 10 |
|---|---|---|---|---|---|---|---|---|---|---|
| 0·1 | 5·0 | | | | | | | | | |
| 0·2 | 9·5 | 1·0 | | | | | | | | |
| 0·3 | 13·7 | 1·4 | | | | | | | | |
| 0·4 | 17·5 | 2·0 | | | | | | | | |
| 0·5 | 21·4 | 3·5 | | | | | | | | |
| 0·6 | 24·8 | 4·6 | | | | | | | | |
| 0·7 | 28·1 | 6·0 | 1·2 | | | Less than 1 per cent | | | | |
| 0·8 | 31·1 | 7·2 | 1·5 | | | | | | | |
| 0·9 | 35·2 | 8·9 | 2·0 | | | | | | | |
| 1·0 | 36·8 | 10·6 | 2·4 | | | | | | | |
| 1·2 | 41·8 | 13·6 | 4·3 | 1·0 | | | | | | |
| 1·4 | 46·2 | 17·0 | 5·2 | 1·4 | | | | | | |
| 1·6 | 50·0 | 20·4 | 6·9 | 2·1 | | | | | | |
| 1·8 | 53·6 | 23·6 | 8·7 | 2·6 | | | | | | |
| 2·0 | 56·8 | 27·0 | 10·8 | 3·5 | | | | | | |
| 2·5 | 63·3 | 34·8 | 16·6 | 6·0 | 1·7 | | | | | |
| 3·0 | 68·3 | 41·7 | 22·5 | 10·9 | 4·6 | 1·7 | | | | |
| 3·5 | 72·3 | 47·6 | 28·2 | 15·0 | 7·1 | 3·0 | 1·2 | | | |
| 4·0 | 75·5 | 52·7 | 33·7 | 19·5 | 10·2 | 4·8 | 2·1 | | | |
| 4·5 | 78·0 | 57·2 | 38·8 | 24·2 | 13·8 | 7·1 | 3·4 | 1·4 | | |
| 5·0 | 80·1 | 61·0 | 43·4 | 28·7 | 17·6 | 9·9 | 5·1 | 2·9 | 1·5 | |
| 5·5 | 81·9 | 64·2 | 47·6 | 32·9 | 21·4 | 12·9 | 7·3 | 4·1 | 2·2 | |
| 6·0 | 83·4 | 67·0 | 51·4 | 36·9 | 25·4 | 16·2 | 9·7 | 5·6 | 3·1 | 1·9 |
| 6·5 | 84·7 | 69·4 | 54·7 | 40·7 | 29·1 | 19·2 | 12·1 | 7·1 | 4·1 | 2·2 |
| 7·0 | 85·7 | 71·5 | 57·7 | 44·6 | 32·8 | 22·8 | 14·9 | 9·2 | 5·4 | 2·9 |
| 7·5 | 86·7 | 73·4 | 60·2 | 47·6 | 35·4 | 25·9 | 17·5 | 11·5 | 6·9 | 4·5 |
| 8·0 | 87·5 | 75·0 | 62·7 | 50·7 | 38·5 | 29·4 | 20·9 | 14·3 | 9·6 | 6·6 |
| 9·0 | 88·9 | 77·8 | 66·8 | 55·9 | 45·4 | 35·5 | 26·7 | 19·2 | 13·2 | 8·6 |
| 10·0 | 90·0 | 80·0 | 70·0 | 60·1 | 50·4 | 41·1 | 32·4 | 24·6 | 17·9 | 12·3 |

Average spare bucket space (records)

**Figure 7.7** *Percentages of inserted records overflowing*

during processing runs, but also to assign them to buckets when the file is being created. This means that there is the possibility of overflow occurring during file creation, because although the algorithm might be perfect for an even spread of keys, in reality this is not often the case. It can be shown from probability theory that the expected number of buckets $b_r$ that will have $r$ records assigned to them is given by the formula:

$$b_r = \frac{B \exp[-R/B]}{r!} \left[\frac{R}{B}\right]^r$$

where $B$ is the total number of buckets in the file, $R$ the total number of records in the file.

This is made clearer by using an example. Suppose there are 20,000 records to be stored randomly in 4000 buckets, then

$$R/B = 20,000/4000 = 5$$

and

$$B \exp[-R/B] = 4000 \times 2.718^{-5} = 26.96$$

Thus, for instance, the number of buckets with two records assigned is

$$26.96 \times 5^2/2! = 337$$

In a similar way, the number of buckets with other numbers of records assigned can be calculated and put in the form of a table as shown in Figure 7.8. These figures have been rounded off to whole numbers which accounts for the small discrepancies in the totals.

Suppose the buckets are large enough to accommodate up to 7

| Number of buckets | Number of records assigned | Cumulative total buckets | Cumulative total records |
|---|---|---|---|
| 27 | 0 | 27 | 0 |
| 135 | 1 | 162 | 135 |
| 337 | 2 | 499 | 809 |
| 562 | 3 | 1,061 | 2,495 |
| 702 | 4 | 1,763 | 5,303 |
| 702 | 5 | 2,465 | 8,813 |
| 585 | 6 | 3,050 | 12,323 |
| 418 | 7 | 3,468 | 15,249 |
| 261 | 8 | 3,729 | 17,337 |
| 145 | 9 | 3,874 | 18,642 |
| 73 | 10 | 3,947 | 19,372 |
| 33 | 11 | 3,980 | 19,735 |
| 14 | 12 | 3,994 | 19,903 |
| 5 | 13 | 3,999 | 19,968 |
| 2 | 14 | 4,001 | 19,996 |
| 1 | 15 | 4,002 | 20,011 |

Figure 7.8 Records assigned to buckets

average records; this corresponds to an approximate packing density of

20,000/4000 × 7 = 71 per cent.

We can now calculate the number of records overflowing. Looking at the cumulative totals at the 7 records assigned level, we see that the first 3468 buckets have 15,249 records assigned to them. This means that the remaining 532 buckets are all full because an attempt is made to assign 8 or more records to each one, thus they hold 7 records each, i.e. 532 × 7 = 3,724 records between them. The total number of records stored in their home buckets is 15,249 + 3724 = 18,973, leaving 1027 to be stored in overflow buckets, i.e. 5.1 per cent.

This figure can be confirmed by interpolating in Figure 7.9; this shows the percentage overflows for various packing densities and bucket sizes.

| Bucket size in records | Packing density = $\dfrac{\text{Number of records}}{\text{Storage capacity in records}}$ | | | | | |
|---|---|---|---|---|---|---|
| | 50% | 60% | 70% | 80% | 90% | 100% |
| 1 | 21·3 | 24·8 | 28·1 | 31·2 | 34·1 | 36·8 |
| 2 | 10·4 | 13·7 | 17·0 | 20·4 | 23·8 | 27·1 |
| 3 | 6·0 | 8·8 | 12·0 | 15·4 | 18·9 | 22·4 |
| 4 | 3·8 | 6·2 | 9·1 | 12·3 | 15·9 | 19·5 |
| 5 | 2·5 | 4·5 | 7·1 | 10·3 | 13·8 | 17·6 |
| 6 | 1·7 | 3·4 | 5·8 | 8·8 | 12·2 | 16·1 |
| 7 | 1·2 | 2·6 | 4·7 | 7·6 | 11·0 | 14·9 |
| 8 | 0·8 | 2·0 | 4·0 | 6·7 | 10·1 | 14·0 |
| 9 | 0·6 | 1·6 | 3·4 | 5·9 | 9·3 | 13·2 |
| 10 | 0·4 | 1·3 | 2·9 | 5·3 | 8·6 | 12·5 |
| 12 | 0·2 | 0·9 | 2·2 | 4·4 | 7·5 | 11·4 |
| 14 | 0·1 | 0·6 | 1·7 | 3·6 | 6·7 | 10·6 |
| 16 | 0·1 | 0·4 | 1·3 | 3·1 | 6·0 | 9·9 |
| 18 | 0·1 | 0·3 | 1·0 | 2·7 | 5·5 | 9·4 |
| 20 | | 0·2 | 0·8 | 2·3 | 5·0 | 8·9 |
| 25 | | 0·1 | 0·5 | 1·7 | 4·1 | 8·0 |
| 30 | | | 0·3 | 1·2 | 3·5 | 7·3 |
| 40 | | | 0·1 | 0·7 | 2·6 | 6·3 |
| 50 | | | 0·1 | 0·5 | 2·0 | 5·6 |
| 70 | | | | 0·2 | 1·4 | 4·8 |
| 100 | | | | 0·1 | 0·8 | 4·0 |

**Figure 7.9**  *Percentages of initial records overflowing in a random file*

## 7.5    Security of direct access files

It is an unfortunate fact of life that no equipment, computer programs, nor computer staff are infallible. This means that every
system must incorporate measures to ensure that no data is permanently lost. This safeguard is most important for all for master
files because these may have been updated and amended many
hundreds of times since they were originally created, and it is therefore out of the question to re-create them from the original source
data.

The ways in which master file data can be lost from direct access
files are mainly as follows:

1    Incorrect transaction data is used to update the master file. This
     should obviously be detected by the system before damage is
     done.
2    A programming or operational error results in the accidental
     overwriting of master data; very probably the overwriting data
     has no connection whatsoever with the master data.
3    Damage to removable disks during use, storage or transit; this
     might be physical or electrical (magnetic) damage.
4    Breakdown or temporary failure of equipment; the latter could
     be caused by mains interference, power failure, or unsuppressed
     radiation from other equipment.

With sophisticated systems the recovery procedure after the detection of an error is automatic and independent of the particular
program(s) in operation at the time. This must be so with real-time
and multi-programming systems, because the operators cannot be
expected to manually trace faults in such complex systems. With
more mundane systems, however, file security has to be arranged by
the user, and a few methods are described below.

*File copying*

This involves copying (dumping) updated direct access files on to
magnetic tape or another area of direct access storage. The precise
method is dictated by the configuration of the computer and in some
cases, especially with fixed disk storage devices, magnetic tape is included mainly for security purposes. When doing a long updating run,
the runs are split into sections, and a dump made after the completion of each section. This enables a re-start to be made at the beginning of the whole file.

*After-state copying*

During an updating or amendment run, this method calls for the buckets that have had changes made to a record therein to be written to another file. This occurs every time a bucket's records are changed, so that at the end of the run or after several runs, there are several 'after-state' versions of each bucket. These are then sorted and amalgamated to produce the most recent version of each bucket, and these are merged into the original file to produce the updated file.

A master file is thus overlayed far less frequently than usual, with a corresponding reduction in accidents. The method is suitable for real-time and on-line applications where the transactions and amendments are not always preserved.

*Record deletion*

Upon occasions some of the records in a file become obsolete and serve no purpose by remaining in the file. Once the user is certain that no further information is extractable from them, they can be deleted from the file. In some applications this is an automatic process, for example where an address file of transient customers is in use. If the date is inserted into the record by the computer every time it looks up the address, it becomes a straightforward procedure for a housekeeping run to delete records that have not been referred to for over, say, one year.

As a general principle, a printed copy should be made of every record deleted from a file. This serves two purposes, first the copy can be retained for possible reference, second it is used to check that all the intended records have been deleted, and no others. This check is especially necessary when the deletions are based on human decisions. Each record will have been found and deleted by the use of a key, such as the customer's account number in an address file, and it is vital that there has been no mistake in the key used; by printing the name and address a further human check can be made.

An extension of this type of check is for the computer not to delete an obsolete record immediately, but to insert a deletion marker into it when printing the hard copy. During the next housekeeping run, all marked records are automatically deleted. If in the meantime an error is found to have occurred, the deletion marker is removed before the housekeeping run occurs. This method avoids having to re-insert records that have been deleted erroneously.

## 7.6 Timing direct access processing

It is relatively straightforward to estimate the time to process a direct access file provided the quantitative details of the file and the access and transfer times of the device are known. Using the formulae given

| Transactions | Cylinders | | | | | | | | | |
| --- | --- | --- | --- | --- | --- | --- | --- | --- | --- | --- |
| | 100 | 200 | 300 | 400 | 500 | 600 | 700 | 800 | 900 | 1000 |
| 1,000 | 100 | 199 | 289 | 367 | 432 | 486 | 532 | 570 | 603 | 632 |
| 2,000 | 100 | 200 | 299 | 397 | 491 | 579 | 660 | 734 | 802 | 865 |
| 5,000 | 100 | 200 | 300 | 400 | 500 | 600 | 699 | 798 | 896 | 993 |
| 10,000 & over | 100 | 200 | 300 | 400 | 500 | 600 | 700 | 800 | 900 | 1000 |

**Figure 7.10**  *Cylinders hit*

below and the tables in Figures 7.10 and 7.11, the processing times can be approximated. These times are, of course, additional to those taken by peripheral devices, taking into account the particular computer's degree of simultaneity. They are also somewhat confused by the operating system but nevertheless are relevant for large-scale processing.

*Sequential processing*

This refers to sequential access to a sequentially stored file. All the buckets are read into core store and those holding records that are updated are written back to the direct access storage device.

If $S$ is the number of cylinders in the file, $T_s$ (msec) the minimum time to access a cylinder (this is used because adjacent cylinders are accessed in turn), $B$ the number of buckets in the file, $b$ the number of buckets hit (from Figure 7.11), $T_r$ (msec) the time to access and read a bucket and $T_w$ (msec) the time to access and write a bucket, then the file processing time is:

$$(ST_s + BT_r + bT_w)/60,000 \text{ min}$$

Take as an example the updating of a file of 50,000 records occupying 10,000 buckets in 200 cylinders, with 30,000 transactions. From Figure 7.11 we find that the number of buckets hit is 9502, and from Figure 7.10 that all the cylinders are hit. Assuming the device has a minimum cylinder access time ($T_s$) of 30 msec, a bucket access plus read time ($T_r$) of 40 msec, and a bucket access plus write time ($T_w$) of 70 msec, the file processing time is:

$$[(200 \times 30) + (10,000 \times 40) + (9502 \times 70)]/60,000 = 17.9 \text{ min}$$

119

| BUCKET IN FILE | TRANSACTIONS | | | | | | | |
|---|---|---|---|---|---|---|---|---|
| | 1,000 | 2,000 | 5,000 | 10,000 | 20,000 | 30,000 | 50,000 | 100,000 |
| 1,000 | 632 | 865 | 993 | 1,000 | 1,000 | 1,000 | 1,000 | 1,000 |
| 2,000 | 787 | 1,264 | 1,836 | 1,987 | 2,000 | 2,000 | 2,000 | 2,000 |
| 3,000 | 850 | 1,460 | 2,433 | 2,893 | 2,996 | 3,000 | 3,000 | 3,000 |
| 4,000 | 885 | 1,574 | 2,854 | 3,672 | 3,973 | 3,998 | 4,000 | 4,000 |
| 5,000 | 906 | 1,648 | 3,161 | 4,323 | 4,908 | 4,988 | 5,000 | 5,000 |
| 6,000 | 921 | 1,701 | 3,392 | 4,867 | 5,786 | 5,960 | 5,999 | 6,000 |
| 8,000 | 940 | 1,770 | 3,718 | 5,708 | 7,343 | 7,812 | 7,985 | 8,000 |
| 10,000 | 952 | 1,813 | 3,935 | 6,321 | 8,647 | 9,502 | 9,933 | 8,000 |
| 12,000 | 959 | 1,842 | 4,089 | 6,785 | 9,733 | 11,015 | 11,814 | 11,997 |
| 15,000 | 967 | 1,872 | 4,252 | 7,299 | 11,046 | 12,970 | 14,465 | 14,981 |
| 20,000 | 975 | 1,903 | 4,424 | 7,869 | 12,642 | 15,537 | 18,358 | 19,865 |
| 30,000 | 984 | 1,935 | 4,606 | 8,504 | 14,597 | 18,964 | 24,334 | 28,930 |
| 40,000 | 988 | 1,951 | 4,700 | 8,848 | 15,739 | 21,105 | 28,540 | 36,717 |
| 50,000 | 990 | 1,961 | 4,758 | 9,063 | 16,484 | 22,559 | 31,606 | 43,233 |
| 60,000 | 992 | 1,967 | 4,797 | 9,211 | 17,008 | 23,608 | 33,924 | 48,667 |
| 80,000 | 994 | 1,975 | 4,847 | 9,400 | 17,696 | 25,017 | 37,179 | 57,080 |
| 100,000 | 995 | 1,980 | 4,877 | 9,516 | 18,127 | 25,918 | 39,347 | 63,212 |

**Figure 7.11** *Buckets hit*

120

## Selective-sequential processing

This corresponds to selective-sequential access and indexed sequential storage. Only buckets holding records that are updated are read and written, and the number of cylinders hit tends to be less than with sequential processing.

If $s$ is the number of cylinders hit (from Figure 7.10) and $T_s$ the minimum cylinder access time|(should be slightly greater than this but the difference is negligible), the file processing time is:

$$[sT_s + b(T_r + T_w)] /60,000 \text{ min plus index search time}$$

Taking the previous example and referring again to Figure 7.10, we find that 200 cylinders are hit and the processing time (excluding index searching) is:

$$[(20 \times 30) + 9502 (40 + 70)]/60,000 = 17.5 \text{ min}$$

## Random processing

This appertains to random access to a randomly or a sequentially stored file. Only the buckets holding updated records are read and written, although any one bucket may be accessed several times because the movements are in random order.

If $m$ is the number of transactions and $n$ the number of cylinder accesses, then,

$$n = 1 + m [1 - (1/s)] \text{ approx.}$$

and if $T_a$ is the effective cylinder access time (based on an average seek distance of $S/3$ cylinders [7.7], the file processing time is:

$$[nT_a + m(T_r + T_w)]/60,000 \text{ min (plus AGA time)}$$

The previous example gives

$$n = 1 + 30,000 [1-(1/200)] = 29,851$$

and the processing time (excluding AGA time and assuming $T_a = 40$ msec) is

$$[(29,851 \times 40) + 30,000 (40 + 70)]/60,000 = 74.9 \text{ min}$$

The times calculated in the above three examples must not be taken as indicative of the relative merits of the three modes of processing for all cases. The selective-sequential and random processing times should be increased slightly to allow for the index retrieval and searching time or the algorithm time. The sequential and selective-sequential times do not, of course, include the times taken to sort the transactions into the same sequence as the file; this may be large.

The timing formulae described above also do not include any allowance for the processing of overflow records. This extra time depends upon a number of factors, predominant among which is the number of buffer areas available in core store to receive the buckets. If two buffer areas are available, it is then possible with a sequential file to hold a home bucket and the corresponding overflow bucket in core store simultaneously, and thereby reduce the overflow processing time. In general, the effect of overflow is to increase the processing time by not more than 10 per cent.

## 7.7 References and further reading

7.1 'Integrity of a mass storage filing system', *Computer Journal* (February 1969).
7.2 'Recovery procedures for direct access commercial systems', *ibid.* (May 1970).
7.3 'Direct access algorithms', *ibid* (August 1971).
7.4 'Operation of a disk data base', *ibid* (November 1972).
7.5 'Special report on disks', *Data Systems* (May 1972).
7.6 'Analysis of self-indexing, disk files', *Computer Journal* (August 1975).
7.7 'Estimating magnetic disk seeks', *ibid.* (February 1975).
7.8 'Bringing cost into file design decisions', *ibid.* (August 1975).
7.9 'A dynamic disk allocation algorithm designed to reduce fragmentation during file reloading', *ibid.* (November 1971).
7.10 MARTIN, *Computer data-base organization*, Part II, Prentice-Hall (1975).
7.11 'Hit ratios', *Computer Journal* (February 1976).
7.12 'File design fallacies', *ibid.* (February 1972).

# Chapter Eight
# *Magnetic tape*

## 8.1 General characteristics of magnetic tape

Magnetic tape is slowly disappearing as an on-line storage medium, mainly owing to its high access time and the consequent need for repeated sorting runs. It is nowadays used more for the input of edited source data from key-to-disk systems, etc., and also as a reserve storage medium for back-up files and dumped data.

It is worthwhile to list the characteristics of magnetic tape so as to have a basis of comparison with direct access devices:

1   A reel of magnetic tape can hold several million characters of stored data.
2   Data is transferred to or from the tape at speeds of 10,000 to 300,000 characters per second.
3   Data can be stored indefinitely or erased and new data recorded on the same reel. Since reels of magnetic tape are comparatively cheap as compared with magnetic disks, it is feasible to hold a large amount of data off-line in the tape library.
4   The length and content of data records are extremely flexible, more so than with direct access storage.

The main disadvantage of magnetic tape is its inherent characteristic of serial storage. This obviously obstructs its ability to provide immediate access to stored data, with consequent problems in realtime and interactive systems.

## 8.2 Records and blocks

The electromechanical principles of magnetic tape units are well known, and there is nothing to be gained from recapitulating them here. We are, however, concerned with the layout of magnetic tape, and this means that it is helpful to appreciate the reason for blocked data. Magnetic tape can be read or written only while it is moving and since data cannot be processed at the same time as it is being transferred, a limit is imposed on the amount that is dealt with in one transfer. This is a 'block' of data, and is a physical concept in that it is transferred automatically by the computer. At the end of a block the computer ceases to transfer data because an 'end of block marker' is detected. The size of the block within a file is decided at the time the file is created, and thereafter remains constant. The main deciding factor is the amount of main store that can be spared for holding the block (the buffer area). Several buffer areas may be needed if a number of magnetic tape units are employed simultaneously. After the cessation of data transfer at the end of a block, the tape continues to move so that the read/write head becomes adjacent to the 'inter-block gap' in which no data is recorded. The size of the inter-block gap is decided initially by the rate at which blocks can be made ready in core store for transfer to tape. It is important that the block to gap length ratio is as high as possible as this affects the time taken to process the tape. The inter-block gap has a maximum length, so that in many cases it is the block length that is the deciding factor. A very considerable difference in read/write time is experienced with variation in block length; for example, with a given tape speed and packing density, it is quite possible for a change in block length from 100 to 1000 characters to reduce the reading time from 9.4 to 2.4 min for the same number of records. Actual block lengths used vary from one computer and from one application to another, but in round terms they lie between 100 and 2000 characters.

The magnetic tape record is a logical concept, its size being related solely to its application usage, and in no way controlled by the hardware. Thus a record may be smaller or greater in length than a block, although in practice it is usually smaller. The relationship of blocks to records is completely flexible and even within the one file is not necessarily constant. As has been explained earlier (Section 6.2), record lengths may be variable due to their holding either *(a)* a variable number of data items, or *(b)* data items of variable length, or *(c)* both of these. In the extreme case where data items, records and blocks are all variable, it is necessary to employ various 'markers' so as to indicate their end points.


## 8.3 | File updating and amendment

The same principles of updating and amendment apply to magnetic tape files as for direct access files. Similarly, magnetic tape files can be classified according to their contents in the same way as direct access files, that is into transaction, transition and master files.

*File — reel relationship*

The fact that magnetic tape is in limited lengths, i.e. reels, is purely for the convenience of handling it. Reels are up to 3600 feet in length and this length of tape can be accommodated on one spool.

There is no fixed relationship between a reel and a file. It is quite normal for long files to occupy several reels, called 'multi-reel files'. On the other hand, it is often convenient to hold several smaller files on one reel, called 'multi-file reels'. These arrangements need cause no difficulty provided their organization is efficient and the magnetic recording of labels by the software is implemented. Nevertheless it is often convenient to have only one file per reel, if necessary using shorter reels so as not to waste tape. This arrangement is unavoidable when files have to be processed concurrently in the same computer run.

*Updating and amendment runs*

The only viable method for comparing records on two or more magnetic tape files is by first sorting them into the same sequence. Thus when using a transaction file to update or amend a master file, it must be sorted into the same sequence as the master file. For some computer runs it is also necessary to sort the transactions into a secondary sequence within the primary sequence. This applies, for example, to stock updating and allocation. In this application some movements have priority and therefore the sort must be into date sequence within stock number sequence.

*Brought-forward/carried-forward files method* Because master file records cannot be rewritten to the same tape reel as they have just been read from, it is necessary to employ two tape decks for handling master file data during an updating or amendment run; this is shown in Figure 8.1. The brought-forward master file is read along with the transaction file, both in the same sequence. When record keys coincide, the master record is updated or amended by the related transactions, and the new version of the record written on to the carried-forward file. Master records that are not changed are copied identically on to the carried-forward file. The carried-forward file from one run becomes the brought-forward file for the next run. When using this method for updating and amending a master file, it is necessary to allow for transactions that do not have a corresponding master record. This condition usually indicates that an error has occurred in some previous stage of the system.

*Changes tape method* This is an alternative to the previous method, using four tape decks instead of three. The principle employed is to read the transactions and the master file as before, but to write only changed master records to a 'changes' tape. During the next updating run, the 'changes' tape, the master file, and the next lot of transactions are all read. Any records from either the changes tape or the master file that are changed are written to another changes tape. Thus after the second week's run, the latest changes tape holds an up-to-date version of every master file record that has been changed during week 1 and/or week 2.

This procedure is shown in Figure 8.2 for an application in which the transactions are read weekly but the master file is updated only every four weeks. Provided it is not necessary to have the master file in an up-to-date condition except at the four-week period end, the method can be used as a means of reducing the over-all processing time. This is especially true if the transactions tend to apply to only a small proportion of the master records (low activity file).

As a rough guide to comparative times, if the volume of movements in each run is about one-tenth of the master records, and the same keys predominate on the changes tape, the changes tape method would take approximately three-quarters of the time of the brought-forward/carried-forward method.

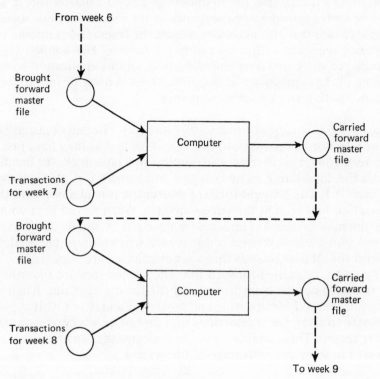

**Figure 8.1**  *Updating a master file*

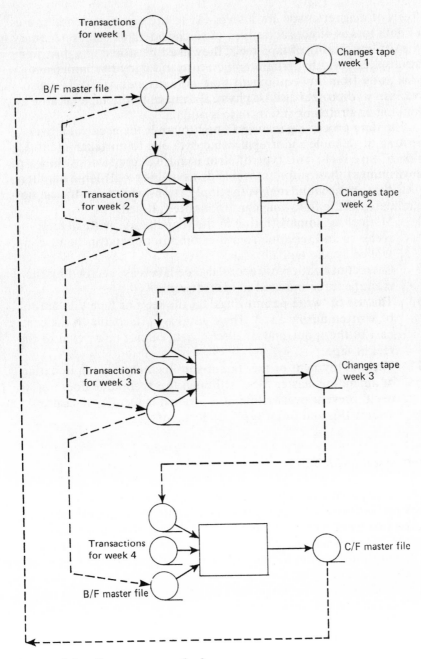

**Figure 8.2** *Changes tape method*

## 8.4 Security of magnetic tape files

Reels of magnetic tape are susceptible to the same accidental causes of data loss as direct access files. Also considerable care is necessary in the off-line storing of tape reels; they must be stored in a dust-free atmosphere, within certain temperature and humidity limits, and well away from any equipment that has a magnetic field. It is also necessary to protect against physical damage to the tape because any kinking or stretching results in loss of data.

The data processing manager's nightmare is the accidental over-writing of valuable and irreplaceable data due to mistaken identification of tape reels. This type of error is more likely to occur in a tape environment than with removable disks owing to the large number of reels that are accumulated as the applications widen. Errors can be eliminated if the following conditions are fulfilled:

1   Meticulous administration in the tape library, with all reel issues to and receipts from the computer operators being controlled by the tape librarian.
2   Correct utilization of recorded tape labels as specified by the manufacturer and used in his software.
3   The use of 'write-permit rings' on the reels of tape that are to be written during a run. These rings are attachable to the centre part of the spool, and, if absent, data cannot be written to the reel of tape.
4   The employment of the 'father-son' technique when updating or amending master files. Although this technique does not, in itself, prevent overwriting, it is a valuable security measure against this and other types of accidents.

*Father-son technique*

The principle behind this technique is simply to retain a brought-forward tape for some time after the creation of the corresponding carried-forward tape. It is customary always to have at least two 'generations' of tape in existence at any time, as is seen from Figure 8.3. Thus during an updating run there are three tapes in temporary existence.

1   The 'son' tape in the process of being created on a fresh reel, i.e. the carried-forward tape.
2   The 'father' tape, which was the previous week's son tape, being read, i.e. the brought-forward tape.
3   The 'grandfather' tape, which was the previous week's father tape, retained on the shelf for security purposes.

When a son tape has been fully created and validated, the grandfather tape becomes eligible for scratching, i.e. re-use for other data. When

128

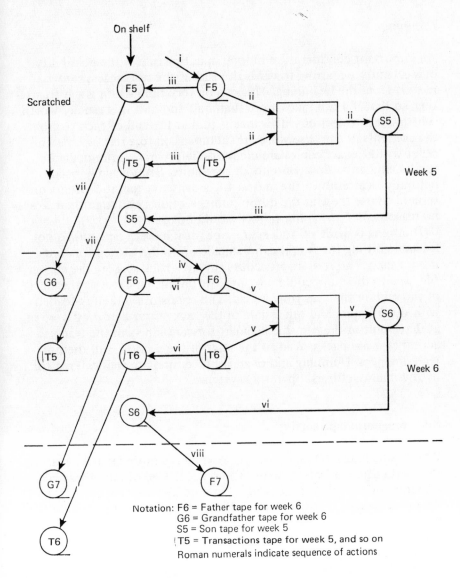

On shelf

Scratched

Week 5

Week 6

Notation: F6 = Father tape for week 6
G6 = Grandfather tape for week 6
S5 = Son tape for week 5
⏐T5 = Transactions tape for week 5, and so on
Roman numerals indicate sequence of actions

**Figure 8.3**   *Father-son technique*

the next updating run starts, the son becomes the father and the
father becomes the grandfather tape. It is necessary to retain also the
transaction tape from the previous run. By adopting these measures
it is possible to re-create a son tape by re-running the previous run's
transaction tape and father tape. This enhances the security of the
system by safeguarding against loss of data through tape damage or
accidental overwriting. A higher degree of security is achieved by
retaining the tapes for more generations, although practical considerations of tape cost and shelf storage space mitigate against this.

129

An important consideration in long updating runs is the possibility of a calamity occurring towards the end of the run, necessitating a re-start from the beginning. This can be very time wasting and frustrating, so that if a run takes more than half an hour, it is usually worth splitting it into sections. This means that at the end of each section, the contents of the core store are 'dumped' on to a reel of tape; in other words, exact copies are made of the program, intermediate results, and any other contents of core store. Sections may occur naturally as a result of the master file's contents, and it is often convenient to use these as the dump-points sections, although there are no reasons why any other points should not be used.

Dumping is made on to a spare tape, or if a spare tape unit is not available during the run, it can be made on to the carried-forward master tape. The re-start procedure involves back spacing the master file tape to the dumped data (if present on this tape), or to the start of the section that has gone astray. The core store is then re-loaded with the program and other data in the exact form that they were in at that point of the run, the brought-forward tape and the transaction tape are backspaced to the start of the section, and the run recommences. Dumping and re-start procedures are well catered for by the manufacturers' operating systems.

## 8.5   Magnetic tape sorting

When using magnetic tape systems, sorting is a more intrinsic feature than with direct access systems. The serial nature of tape files makes it axiomatic that their processing is in sequence, with the result that the processing runs are interspersed by sorting and merging runs. The principal reasons for magnetic tape sorting are the following:

1   To arrange a transaction file in the same sequence as a master file before an updating or amendment run. This is unavoidable unless the volume of transactions or the size of the master file enables either of them to be held completely in core store throughout the run; this situation is highly unlikely.

2   To list or tabulate a master file in a different sequence from that in which it is usually held. This happens, for instance, with a stock file that is normally held in part number sequence but is needed in bin number sequence for stocktaking purposes.

3   To collate items of a similar nature for the purpose of comparison or summarizing. For example, sales records are sorted into area with-

in class of trade sequence so as to facilitate comparison of sales to similar customers in the same locality.

4   To merge together two or more files in order to create one larger file. This occurs frequently during master file creation when the data comes from a number of different sets of source documents.

5   To highlight certain factors by bringing them to the beginning of a file. This reason for sorting differs from the others in that the sorting key could well be a quantity or value rather than an indicative field. An instance of this is the sorting of job costs in reverse sequence of their values, thereby putting the higher costs at the top of the job costing list.

The programs required for carrying out magnetic tape sorting are written by the manufacturer's programmers, and supplied to the user as part of the software suite accompanying the hardware. They are usually in the form of a 'sort generator' program that can create the most suitable type of sorting routine for the job in hand. The sort generator functions by being provided with a set of parameters by the user when he wishes to set up a sorting run. The parameters cover the details of the data to be sorted and the hardware to be used, and include:

1   The number of records to be sorted.
2   The sizes and positions within the records of the fields comprising the sorting key.
3   The degree of initial sequentiality (if any).
4   Forward or reverse sorting indication.
5   The number of tape decks available.
6   The amount of core store available.

## 8.6   References and further reading

8.1   'Magnetic tape cassettes', *Computer Bulletin* (July 1971).
8.2   'Ageing of magnetic tape', *Computer Journal* (November 1968).
8.3   'Curing the magnetic drop-outs', *Data Systems* (November 1970).
8.4   'Magnetic tape-static', *Data Processing* (July-August 1970).
8.5   'Blocking sequentially processed magnetic files', *Computer Journal* (May 1971).
8.6   'Flexibility of block length for magnetic files', *ibid.* (November 1973).
8.7   'Magnetic tape ending', *Data Systems* (March 1973).
8.8   'Special on magnetic tape', *ibid.* (June 1972).
8.9   *Magnetic tape standards*, British Standards BS 3658 (1963), BS 3968 (1966) and BS 4503 (1969).
8.10  'Care of magnetic tape', *Data Processing* (November-Debember 1969).

# Chapter Nine
# *Systems design considerations*

## 9.1   The computer as a service to the organization

Along with most other departments in an organization, the data
processing department is not an end in itself. With the exception of
when, if ever, it is doing chargeable work for outside agencies, the
department does not appear to be directly involved in making money
– nor sometimes in saving money – and its role as a service to the other
service departments does not give it the most exciting of images.

The initial impact of a newly installed computer often draws a
considerable amount of attention towards a new system. It is part of
the systems analyst's function to ensure that this interest in maintained at a sensible level thereafter; the computer's opening role as
the centre of attention must not be followed by a lifetime of remote
isolation. What can the systems analyst do when designing the data
processing system in order to make the user department staff feel
that they are part-owners of the computer?

*Encourage communication*

Whenever suitable the new system should be designed to allow
amendments and inquiries to be accepted regularly from the user
departments. This does not imply that the departments should be
encouraged to request special amendments in a never-ending stream,
but that the ability to cater for pre-planned amendments is built
into the system. This means that amendments can be dealt with
automatically without the need to write unforeseen programs or
perform special computer runs.

Nevertheless the need for genuine inquiries must be ascertained, and methods for coping with them included in the system. Every effort should be made to dispel the aura of remoteness that can quickly become associated with data processing.

## Encourage participation

In connection with inquiries and other requirements, arrangements should be made for the user departments' staff to obtain their needs from the computer automatically and without making special requests. This can be done in two main ways.

1   By the submission of standard inquiry forms to the data processing department. These inquiries would be dealt with as a matter of routine and at predetermined times, most probably when the relevant file was already available to the computer.

2   By the writing of enquiry programs by the user departments' staff, probably in COBOL, FORTRAN or other high-level language. This arrangement is particularly efficacious for meeting the needs of technical staff such as design engineers and research workers. In cases where the computer is to be used for relatively small amounts of technical work, it is easier for the technical man to learn the program in say, FORTRAN, than for the systems analyst to learn the technical work.

It is unlikely that user department staff can write successful programs for other than basic enquiries using a terminal.

## Arrange 'on-demand' output

As applied to data processing, the principle of management by exception has many advantages and should therefore be implemented whenever possible. There are, however, some situations in which a large volume of printed output is unavoidable. These are generally where there is a need either for readily available reference information in the form of a directory, or for special reports in order to analyse mentally the information therein. A danger here is that reference information swamps the recipient and, as a consequence, is seldom used. Similarly, special 'once-only' reports become perpetuated, resulting in an ever-increasing amount of printed matter being prepared by the computer.

A solution is to make these types of computer output into 'on-demand' items. The applicants for special reports must then submit a request on a standard form. As a reminder of this service, and a guide to the latest dates in the week, month, etc., by which requests should be submitted, other regular computer output can have appended details printed by the computer. If the special reports occupy little computer processing time, it may be worth preparing them regularly but distributing only to persons making specific requests on that occasion.

## 9.2 Inter-relationship of files

It was suggested in an earlier chapter that the files are the skeleton of a data processing system. Just as a skeleton has to fit together in a flexible manner, so must files in order that they can perform their function. This concept applies to all classifications of files, and most of all to master files because these are the backbone of the system.

A paradoxical feature of systems design is that a file's contents cannot be decided exactly until all the related applications have been designed, and yet no application is fully designed until its files are laid out precisely. This problem is overcome by iterative methods of design by which the file's contents and layouts are decided gradually, being inserted as each application is designed.

Alternatively, the utilisation of a database is program-independent and this means that it is also application-independent.

When designing a logical files layout and contents, the systems analyst has to bear in mind the following points:

1   The contents of a file must be adequate to meet the requirements of all routines that will use it; these may include routines that will be implemented at a later date.

2   If there is doubt as to the certainty of a data item being required in the records in a file, it is usually simpler to include it when the file is created than to have to insert it later. A possible disadvantage in this is the perpetual updating or amendment of an unused data item. An alternative is to leave space for it, and although this is wasteful of storage, it is generally preferable to having to re-organize the whole file later.

3   Separate files that are to be sorted together must have their keys in the same relative position within the records. This ruling sometimes also applies to matching and merging programs supplied by the manufacturer.

4   With smaller computers the magnetic tape block length or the bucket size of a direct access storage device should be planned so as to be suitable for all computer runs that use the file. This factor

# FILE SPECIFICATION

| File name | Ref. | Classifications (Delete as necessary) | | Normal sequence | Created by Run ref. |
|---|---|---|---|---|---|
| Product master cost file | T2 | Transaction ~~Transition~~ Master | ~~Serial~~ Sequential ~~Random~~ | Product code | A3. |

| Ref. | Data item name | Picture | Character positions | Remarks |
|---|---|---|---|---|
| 1 | Product code | 99999 | 1 - 5 | |
| 2 | Std. labour price | 9999.99 | 6 - 11 | In pence + hundredths |
| 3 | Std. material price | 9999.99 | 12 - 17 | "       "       " |
| 4 | Max. discount % | 99.999 | 18 - 22 | Allows for fractions |
| 5 | Selling price | 99.99 | 23 - 26 | |
| 6 | Description | x (20) | 27 - 46 | |
| 7 | Manufacturing group | 9 | 47 | |
| 8 | Batch size | 999 | 48 - 50 | |
| 9 | Factory no. | 9 | 51 | |
| 10 | | | | |
| 11 | | | | |
| 12 | | | | |

# FILE UTILISATION

| Routine code | Routine name | Data items used — refs | | | | | | | | | | | |
|---|---|---|---|---|---|---|---|---|---|---|---|---|---|
| | | 1 | 2 | 3 | 4 | 5 | 6 | 7 | 8 | 9 | 10 | 11 | 12 |
| E | Order handling | ✓ | | | | ✓ | ✓ | ✓ | ✓ | ✓ | | | |
| G | Production planning | ✓ | | | | | | ✓ | ✓ | ✓ | | | |
| H | Invoicing | ✓ | | | | ✓ | ✓ | | | | | | |
| K | Sales analysis | ✓ | ✓ | ✓ | ✓ | ✓ | | ✓ | | | | | |
| M | Production control | ✓ | | | | | | ✓ | ✓ | ✓ | | | |
| P | Sales forecasting | ✓ | ✓ | ✓ | ✓ | ✓ | ✓ | | | | | | |
| T | Costing | ✓ | ✓ | ✓ | | ✓ | | ✓ | | ✓ | | | |

**Figure 9.1**   *File specification*

hinges on the core store taken by the most complex programs in the system. The analyst cannot be expected to know the exact program size at the systems design stage, but approximations can be made so that block or bucket sizes can be decided.

5   Upon occasions, files contain records or data items that are not actually used in the regular data processing routines. These insertions are often for reference purposes in connection with inquiry and interrogation procedures. In other cases their presence is as a security precaution, such as when a set of entities has recently been re-coded and the old code numbers are inserted in order to remove any possibility of confusion.

*File record usage*

As a means of deciding the format and ensuring the completeness of file records, it is helpful to fill in a 'file specification sheet' for each file that is to be created (Fig. 9.1). This document shows the contents of the file's record(s) and its utilization by the processing routines. Each field is specified and cross-reference to the file utilization, in which it is ticked off against the routines that make use of it. In this way the chance of a field being accidentally omitted from a file is minimized. In systems where the routines consist of a large number or a complex arrangement of computer runs, the fields could be ticked off against the runs instead of the routines. The precise method of specifying the records depends upon the storage device employed and the characteristics of the computer's addressing system; Fig. 9.1, however, gives a general indication of the basic requirements.

## 9.3   Integrated data processing

The concept of integrated data processing has been with us ever since the first business applications were applied to computers. It has however, in many cases proved easier to conceive than to translate into a working system. As suggested in Section 9.2, a data processing system is heavily dependent on the accuracy and completeness of its logical files. This need for efficient files was often the stumbling block in moving towards integration. The software needed to control a number of interrelated files is too complicated for most computer users to program themselves. Now that database software (Section 6.7) is available from computer manufacturers and software houses, most organizations are able to achieve a higher degree of integration. The creation of an organization's database makes it apparent that integration is meaningful and to some extent unavoidable. What then is meant by integration within the framework of data processing?

It is often believed by people in the computing world that integration means 'doing everything by computer' — forcing every application into the 'total system'. The nebulous assumption supporting this idea is that the more applications are 'computerized', the more profitable they become since the computer is available at no extra cost. This assumption is occasionally correct, but the systems analyst must, if necessary, disabuse himself from the belief that it is invariably so.

Instead of regarding integration as the total system, it is wiser to aim at a system that also is as 'open-ended' as possible. The latter characteristic implies that further applications can be built into the system with the minimum of difficulty. The main aim of integration should be the uniting of existing data processing applications; attempting to encompass every existing application is not necessarily possible nor desirable.

We must also guard against assuming that integration necessarily entails the employment of a large computer, or for that matter, of a computer at all. In principle a manual system can be fully integrated; in practice this is difficult due to the slowness of manual operations and the problems of high-speed communication between departments. The size of the computer required depends on the volume of its work and files; the degree of integration acquired depends upon the skill of the systems analyst.

### Design features of integration

When designing an integrated system, the analyst is well advised to bear in mind certain basic principles, as outlined below.

*Source data*  This enters the system once only and is thereafter processed within the system as often as is necessary to provide the required output. A typical example is a works job ticket that enters the system once only from the factory; the data from it is then processed to provide information for the payroll, job costing, work-in-progress control, and so on.

*Phasing of routines*  Each routine in the system has a phase relationship with other routines — there are obvious cases where one routine simply cannot function until another has reached a certain stage. In other cases the efficiency of a routine is jeopardized if the related routines get out of phase with it. An example of this is when a stock file fails to be updated with receipts into stock before allocations are made from it. The consequence of this situation is that items are incorrectly designated as out of stock, and sales orders are lost, or the dispatch of goods is delayed.

*Inter-related files* — as described in Section 9.2.

*Common coding*   The efficiency of code numbers depends upon their uniqueness, and it is therefore vital that there is no ambiguity in their meaning between different applications.

*Output reports*   The printing of common information by the computer for use by various departments should be coordinated. Due to the computer's ability to process data in a short time, output reports are more up to date than from other systems. This may result in what appears to be contradictory information on them; and as a safeguard against unjustified criticism, the information may be either time-dated or produced at one and the same time in the form of a combined report.

## 9.4   Distributed processing

A distributed data processing system is one with several interconnected points at which processing power, i.e. intelligence, and storage capacity are available. These points may on occasions act autonomously|and at other times cooperate in handling a common problem. The locations of the processing points need not necessarily be physically remote from one another or from a central mainframe computer.

The main purpose of distributed processing is to give the end users of computing facilities the control over and responsibility for their own data. In some ways this philosophy is a return to pre-computer days. The end user becomes master of his own destiny to a much greater degree than with batch processing carried out entirely within the data processing department. In other words, he had considerable computing power under his control rather than delegating it all to the centralized computer.

Distributed processing presupposes that the user department automatically accepts responsibility for the correctness and completeness of its source data. The department in question is very likely the only body aware of the source data in use and of the immediate results required from the system. Because the source data as such is not handled by the data processing department, its staff cannot be regarded as in any way responsible for errors or for making corrections except when specifically requested by the user department.

138

## Distributed configurations

A distributed processing system may be composed of several processing points connected together in a wide variety of configurations.
The points themselves can be minicomputers, intelligent terminals or
mainframe computers. And with the extending use of microprocessors
the distinction between these is likely to become increasingly blurred.
Microprocessors are now so small and inexpensive that it is economic
to build them into all types of terminals and peripherals.

Broadly, distributed processing systems fall into two approaches—
hierarchical and lateral.

*Hierarchical systems*   These have several levels, the most powerful of
which consists of one or more mainframe computers forming the
central complex of the hierarchy. This complex is capable of handling
local batch processing, remote job entry, time-sharing, and the needs
of the lower levels in the hierarchy. It is likely that the central complex is large and expensive.

The second level comprises a powerful minicomputer(s) acting as a
satellite to the mainframe(s). This must be capable of administering
a network protocol so that data and messages can be passed through
it between the lower and higher levels. This minicomputer must also
be able to handle local batchwork, interactive terminals and, possibly,
communication with other minicomputers.

The third level consists of intelligent terminals dedicated to particular tasks such as point-of-sale processing. They are capable of
controlling a number of keyboards and VDUs, and of communicating
with the second-level minicomputers. At the lowest level of the hierarchy are 'dumb' terminals. These have no intelligence and so act
merely as the means of accepting input and displaying output.

*Lateral systems*   These are similar to hierarchical systems except for
the omission of the mainframe(s). The minicomputers in a lateral
system are autonomous but are capable of communicating with one
another. This intercommunication must be flexible in order that
various arrangements can be set up. In some situations the minicomputers cooperate in order to create a more powerful processing
system, in others the communication is merely the interchange of
messages or data.

The cooperation of minicomputers in a lateral system infers that
they act as stand-by and back-up computers for each other. These
requirements necessitate sophisticated software, and consequently
overheads in terms of storage, cost and time must be taken into
consideration.

From the systems design aspect the following criteria need to be taken into consideration in deciding the suitability of distributed processing.

1    Do the user departments need the capability of the straight-forward processing of large amounts of source data with a rapid turn-round?

2    Is it preferable that the source documents remain continuously in the possession of the user departments? This may be advantageous, for instance, where there are a large number of enquiries and amendments re customers' orders.

3    Can the necessary processing be accomplished by access to locally held files, rather than central files? This could be the case if the file records applied only to the department but not if they required to be updated and/or interrogated by the whole organization.

4    Does the output of the mainframe comprise mainly documents or information required locally rather than centrally?

5    Is it advantageous for source data errors and omissions to be detected immediately on input, and is this feasible without access to central files?

6    Is there a security risk in transmitting data over transmission lines? If so, local processing may be preferable to the adoption of encryption techniques before transmission. Alternatively, does a security risk arise through the local staff having access to processing power and file records?

7    Are the user departments so remote that data transmission costs and error rates would be high? Alternatively, would the physical transmission of source documents and/or output documents be too risky or time consuming?

8    Would failure of a central mainframe computer mean a disruption of work in the user department?

9    Would there be a requirement to move data, file records, programs or software from computer to computer if distributed processing was installed? If this is a requirement, a communications network protocol system will be needed in order to control these movements. Similarly, the DBMS will be more complete owing to the likelihood of one computer needing access to records controlled by other computers.

10    Can the user department operate largely autonomously but with occasional transfers of data to or from other computers? For instance, a local minicomputer may be employed for sales invoicing with daily transference of sales ledger data to the central mainframe.

## 9.5   Systems definition and documentation

The system definition embraces the formal detailed description of the data processing system in the form of written documents. It is essential that the system is documented for three main reasons:

1    As a basis for obtaining official top management approval of the system prior to its implementation.

2    As a means of disseminating information about the system to persons who are involved in its implementation and operation.

3    As a reference document for the future, bearing in mind that its readers may be staff who are completely new to both the system and the organization.

The systems definition documents are, for the most part, prepared during or after the design of the system. It is, however, beneficial to have the proposed contents in mind right from the commencement of design, as this will guide the systems analyst in his deliberations. When preparing the definition, its layout should be arranged to suit the above three purposes. The information provided for top management obviously need not be as detailed as that for the programmers; the emphasis in one case is on aims and advantages, and in the other on precise computer run specifications. It is, in any case, unlikely that the run specifications will be ready by the date at which the system is presented for top management approval (Section 12.4). Similarly, the information needed by the staff in 2 above is not always the same. Included in this group are not only the data processing staff but also other departmental staff.

The preparation of the system definition inevitably takes a fair amount of time, so the systems analyst should avoid duplication of effort. Each section of the definition should be written with all its readers in mind, so that there is no rewriting of the same information. By carefully sectionalizing the document, the appropriate sections can be collated for presentation to the particular recipient, omitting the sections that are of no interest to him. No hard and fast rules can be laid down regarding the distribution of sections; the systems analyst must judge each section's pertinence in relation to its possible recipients.

*Amendments to the system definition*   It must be remembered that the system definition is likely to incur amendments, particularly in the more detailed sections. When these occur a memorandum should be sent to the appropriate staff in order to explain them, and if necessary a whole section re-distributed to replace the original. To facilitate the distribution of amendments and new sections, a list of recipients can be included at the start of each section. The problem of amendment distribution is of course another good reason for restricting the distribution of sections to interested parties only.

It is not the intention to describe here each section of a system definition, this is done in other chapters. A list of suggested sections together with a reference to the relevant section of this book follows below.

1     List of contents and amendments record.
2     Distribution list of each section of the system definition.
3     Aims and advantages of the data processing system (Sections 3.1 and 12.1).
4     A general description of the over-all system, avoiding technical jargon but emphasizing departmental staff participation and system security (Sections 7.5, 8.4 and 9.1).
5     The specifications of all equipment involved, and a brief explanatory note of new or unusual items of hardware (Section 11.5).
6     A summary of the estimated costs of the old and new systems (Section 12.1).
7     An implementation time-table (Section 13.6).
8     A schedule of the computer routines (Section 11.3).
9     A flow chart and a description of each computer routine (Section 10.7).
10    Instructions to the user departments (Section 13.4).
11    Data entry instructions (Section 10.3).
12    File specifications (Chapter 6 and Section 13.2).
13    Distribution list of computer output (Section 13.4).
14    Run specifications (Section 10.8), each run being separable for distribution to the appropriate programmer only.
15    A report on the future outlook, and the possible enhancements to the system in relation to the organization's expansion and developments.

## 9.6   References and further reading

9.1   BOUTELL, *Computer oriented business systems*, Chapters 7 and 9, Prentice-Hall (1968).
9.2   'Methodology of computer systems design', *Computer Journal* (February 1974).
9.3   'Planning and control systems in management accounting', *Computer Bulletin* (September 1967).
9.4   'The CAV integrated system', *Data Systems* (September 1967).
9.5   'Design aims and principles for an integrated nominal ledger and costing system', *Computer Bulletin* (June 1970).
9.6   'Systems design—can it be left to the experts?', *Data Processing* (September—October 1970).

9.7   'The profits of computing and systems analysis', *ibid.* (November—December 1968).
9.8   *Documenting systems*, NCC (1969).
9.9   *Systems documentation manual*, NCC (1975).
9.10   *A system documented*, NCC (1974).
9.11   'System documentation—why bother?', *Computer Management* (May 1974).
9.12   LONDON, *Documentation standards*, Petrocelli (1974).
9.13   DOW and TAYLOR, *Why distributed computing?* NCC (1976).
9.14   'The minicomputer', *Data Systems* (March 1975).
9.15   SANDERSON, *Minicomputers*, Newnes-Butterworths (1976).
9.16   MACE, *Visible record computers*, Business Books (1974).
9.17   'Centralize or decentralize?', *Data Processing* (September—October 1970).

# Chapter Ten
# *Design of data processing systems -1*

## 10.1    Stages of systems design

Prior to the start of designing the data processing system, the systems analyst has had comparatively little opportunity to use his creative abilities. As we have seen, his work up to this juncture has, for the most part, been that of observing an existing situation but not attempting to alter it in any way. Throughout the investigation the systems analyst plays with ideas at the back of his mind, and there is no exact time at which he turns from investigation to design. These ideas, perhaps fanciful for the moment, prompt him to seek further information from the existing system so that he can assess the viability of new systems arising from the ideas.

Before turning his efforts to full-time design, it is advisable for the analyst to look afresh at the objectives and at his assignment brief. A reappraisal of these in the light of information now to hand enables him to decide their relevancy and realism. Any doubts in these respects should be eliminated by further discussions with top management before using the objectives and the assignment brief as guidelines.

The task of designing each individual data processing system presents its own special problems and unique features. It is not possible to lay down rigid rules of systems design that apply to all cases. Moreover it is dangerous for a systems analyst to overplay his experience by attempting to impose previous solutions to present problems. Each new situation must be considered against its own background, and the emphasis in the systems design weighted accordingly.

*Aspects of systems design*

When designing a data processing system, the main aspects to be considered are:

1    Its cost.
2    Its efficiency and accuracy.
3    Its practicality.
4    Its flexibility.

The aspects cannot be rated against each other once and for all; in one company the overriding aspect is cost minimization, in another it may be a combination of accuracy and practicality. Even within the one company their relative importance may change with changing circumstances, so that the emphasis given to the aspects in one year might not hold a few years later. Nor can they be treated as isolated and distinct aspects of design, they are irrevocably connected, a change in any one causes some degree of change to each of the others. The weight that the systems analyst gives to each of these aspects must be decided as a result of his knowledge of the company's activities and by an intuitive forecast of future trends.

*Breakdown of systems design — routines and runs*

A data processing system, even though integrated, can be broken down into a number of routines and then further into computer runs. A routine can be regarded as a piece of data processing work that achieves a result that is usable outside the system. A routine forms part or all of an application, and consists of a number of computer runs that are tied together in various ways, and in particular by the output of one run forming the input to another. Examples of routines are payslip printing, sales invoice preparation, product parts explosion, stock updating, and so on. Although a routine can be carried out in segments at separate times, it is usual to carry it through without interruption from start to finish.

A run is a piece of work on the computer that is intended to be carried out as a whole and in a continuous fashion. The extent to which this is achieved depends upon the operating system but is generally a desirable aim. A run is usually, but not always, associated with one computer program which is called into action to control the run. Where two or more runs are interconnected by the output of one forming the input of another, the output is in the form of data on magnetic tape, disks or drums, or occasionally in core store. The output can also be printed documents that are inspected and amended manually before being re-punched or re-read to form the input to another run; this arrangement is a much looser connection, and can also apply to different routines.

*Processing per computer run*   The amount of processing to be aimed at in one run should be as large as possible commensurate with the following conditions:

1   The computer's ability to hold the main program in core store, remembering that the lesser-used segments of program may be held in backing storage (drums or disks) and moved into core when required. With a large computer the main store is sufficient for any single program. And, in any event, the use of virtual storage makes the main store appear to be as large as required by allowing the programmer to address unlimited locations of main store. The virtual storage system automatically reloads the main store with program segments and data from backing storage as and when required.

2   The appropriate peripheral units being available in sufficient numbers; this condition is particularly applicable to magnetic tape decks. The use of an operating system makes this less of a restriction since it can spool data for subsequent input or output, e.g. write output for printing to magnetic tape pending a printer becoming free.

3   The amount of direct access storage needed for files used in the run being available at the one time; this applies especially to removable disks.

4   The files, particularly magnetic tape files, being in the most suitable sequence for the run.

5   Input and output peripherals, especially line printers, not being involved with too many types of data that require different operational handling. For instance, when preparing bills on pre-printed stationery it is inconvenient and time-wasting to keep changing the stationery in order to print intermediate reports or analyses. This again is overcome by an operating system spooling data. It is still a problem with small computers with no operating system or only a rudimentary one.

6   The ability of the run to cope with a reasonable degree of changing circumstances without having to re-design it. There are of course limits to what we can expect of a run in this respect; but what must be avoided are unnecessarily complicated runs that are difficult to follow when amendment is needed.

7   The run's ability to detect and deal with exceptions and error conditions. These may sometimes be merely written to magnetic tape or stored internally for further processing by another run.

8   The run's control and audit procedures being straightforward and complete (Section 11.4).

*Run splitting*   If any of the above conditions cannot be accommod-

146

ated with one run, it must be split into two or more runs. When deciding the point at which to make the split, the following points are worth bearing in mind:

1    When using magnetic tape, it is inefficient to write out large files in the first part of a split run and then to read them again in the same sequence in the second part. This procedure is sometimes unavoidable, but it might be possible to minimize the size of such files by summarizing the data before writing out.
2    Great care must be exercised if data is left in main store from one run to another. This procedure is acceptable provided the two runs are in close succession, and there is never the possibility of a third run intervening at any time owing to operational re-scheduling. This situation does not apply if an operating system is used as this would not allow data to reside permanently in main store.
3    It is often convenient to split a run involving a final print-out at the point just before print-out. The final print-out program is usually unconnected with the previous processing, and can therefore be loaded just prior to use. This method is of real benefit only if the final print-out is of significant complexity.

## 10.2   Punched media as computer input

Punched media comprises mainly 80-column punched cards and, to a lesser extent, 96-column cards and paper tape. These media are steadily disappearing in favour of other methods such as are described in Sections 10.3 and 10.5.

Under certain circumstances punched media has advantages and so brief explanations follow below.

### Punched cards

Punched cards have been in use for many years in one style or other, but the increased use of business computers has resulted in one style becoming dominant. This is the 80-column, 12 position, slotted-hole card. IBM has introduced a smaller card containing 96 columns but this is unlikely to come into extensive use. The most obvious method for preparing cards for computer input is the well-established punching and verifying technique. This is covered in Section 10.3, so we will deal first with the other aspects of punched cards.

*Card pulling files*    Instead of punching manually a batch of cards to represent transactions that occur in the company, cards can be pulled (picked) from a pre-punched 'pulling' file. This method is well proven and is effective for coping with a large amount of source data that could not otherwise be handled in the time available. The pulling file is replenished by reproducing (using a punched card reproducer), manual punching, or re-insertion of the original cards during off-peak work periods. A pulling file contains several cards to represent each entity in the file. Typical examples of the entity sets in pulling files are:

Commodities in a selling range.

Items held in a factory stores.

Constituent parts within manufactured products.

Employees on the payroll.

The number of cards in the pulling file for a given entity depends upon its rate of usage; there must be sufficient to avoid running out before the next replenishment.

*Card per unit files*    An alternative arrangement to the above is for the file to hold cards that represent individual entities actually in stock, this is known as a 'card per unit' file. Fifty cards in the pulling file for an entity means that a quantity of fifty of that entity is actually in the stores. This system is disappearing with the increasing employment of computer files, but it does have the advantage that the stock position is immediately apparent provided an entity is withdrawn from stock only against a card. In situations where orders and inquiries are received by telephone, this arrangement can be the most convenient for maintaining a visible up-to-the-minute representation of the stock position.

*Consumable/non-consumable pulling files*    In order to facilitate pulling, the cards are filed in open racks with labels to indicate the entity's code numbers in the adjacent cards. The cards are generally filed in code number sequence and, as a rule, 'interpreted' along the top edge so that it is easier to read their contents. In addition to the pre-punching of code numbers and other indicative data, it is sometimes possible to have pre-punched quantities in the cards. For each entity, there are a few cards for each quantity that could be involved in a transaction. For example, with an 'orders received' file, if the orders are generally for singles of a certain commodity, then most of this commodity's cards would be pre-punched with '1' in the quantity field. The quantity fields often hold small quantities such as 1, 2, 4, 8 or other groups that predominate in usage. A movement involving an uncommon quantity is catered for by pulling two or more cards.

If larger and more widely varying quantities are in use, the pulled cards are punched manually with the quantity after being pulled. It is also possible to punch other data into the cards after pulling; this means that after use the cards cannot be re-inserted into the pulling file, and this is then a consumable file. When no further data is

punched into the pulled cards, they are generally returned to the
pulling file after use and re-inserted into the correct position therein. This type of file is non-consumable, but of course the cards must
be replaced when they show signs of wear. If a high-speed card reader is in use, it is wise to allow only one reading of the cards, i.e. have
a consumable file.

*Card pulling has the following advantages:*
1    Errors due to the misreading of code numbers and descriptions
     on documents tend to be reduced.
2    Pulling is faster than punching, therefore input data can arrive
     at the computer earlier, or greater volumes can be handled in a
     given time.
3    The punching load is reduced by a non-consumable file, and
     more evenly time-spread if the file is consumable.

*The disadvantages applicable to pulling files are:*
1    The card is inherently tied to one and only one item; this means
     that 'spread' cards (holding several entities, and described later
     in this section) cannot be used. This means, as a consequence,
     that more card reading time is needed on the computer.
2    Card consumption tends to be higher than with spread cards.
3    The pulling file occupies a considerable amount of space, involv-
     ing a large room if a wide range of entities is represented.


*Dual-purpose cards*    These are punched cards that are also used as
documents in that they have handwritten entries made on them; the
entries are then hand punched into the card. If dual-purpose cards are
to be read by a computer after they have been handled in a factory
or office, great care must be taken to ensure that no damage is sus-
tained therein. The damage includes creasing, tearing, fraying and
dirtying; if it is unavoidable the cards should be reproduced and the
new copies read by the computer. A reproducer can accept a far
higher level of damage than can a computer's card reader.

*Spread cards*    These cards hold data appertaining to several entities
usually from 2 to 20. Each entity's data may be completely self-
contained so that, in effect, the spread card is equivalent to a pack
of smaller cards joined together. A spread card may hold fields
applicable to all the entities in the card. It is advantageous to be
able to fill a card in this way because the card-reading speed of a
computer is independent of the card's contents, and it also increases
the manual card-punching speed per entity. Spread cards cannot be
sorted off-line (by a mechanical sorter) into the sequence of the keys,
because of the conflict of the various keys therein.

*Multi-function card machine (MFCM)* This is a computer peripheral (IBM 2560) comprising several units that are capable of executing various punched card operations under computer control. The 80-column cards are moved from the two input hoppers to the five output stackers and, in so doing, a MFCM acts as a card arranging and punching device.

The functions of a MFCM are similar to those of the now-extinct punched card machines such as collators and gang summary punches. The main functions are:

*Reproducing* – punching a new pack of cards as a copy of an old set.

*Gang punching* – data is copied from one card into several others.

*Summary punching* – data from several cards is summarized and punched into one card.

*Sorting* – the cards are brought into a sequence based on a key data item(s) in the cards.

*Merging* – two packs of cards in sequence are combined into one pack in the same sequence.

*Matching* – two packs of cards are compared for a one-to-one similarity to their keys.

*Selecting* – certain cards are picked out from a pack dependent on the values of their keys or other data items.

*Interpreting* – up to six lines of characters or digits are printed across the top of the card to show its punched contents.

*Paper tape*

As with punched cards, paper tape has a history of non-standardization. At one time a multitude of paper tape codes and widths was prevalent, but this has now been pruned to two main types [10.7]. These are the 8-track (or channel) (7 data bits plus parity bit) ISO code, and the 7-track (6 data bits plus parity bit) code, both of which are punched in one-inch wide tape. Most paper tape readers can, in fact, read 5, 6, 7 or 8-track tape, and any tape coding can be translated by sofware into the computer's internal representation.

*Parity bits* In 7 and 8-track paper tape, each character is represented by a set of holes across the tape (a frame) and a means of checking automatically that no hole has been accidentally omitted nor inserted by the punching device is to include a 'parity bit' in each frame. Parity may be either odd or even, but must be consistent within the one reel of tape. With even parity, the parity bit is inserted in order to make the total number of bits, i.e. holes, in each frame into an even number; with odd parity, the total is an odd number.

150

*Sources of paper tape* It is possible to punch data into paper tape by means of key-operated punches and verifiers in a similar way to punched cards (this method is covered in Section 10.3); in addition it can appear as a by-product of other operations. These include accounting machines, cash registers, etc., but in this respect paper tape has been largely superseded by OCR and magnetic media.

## 10.3 Keyed computer input

The two principal keyed methods of data entry are key-to-disk and key-to-diskette. There are several similar methods under various proprietary names but these two cover the main principles of the methodology. Their inherent advantages are causing them to steadily replace punched cards, especially in the larger data processing departments.

### Key-to-disk systems

The main components of a key-to-disk system are a number of key-stations (up to 32) a miniprocessor, a supervisor's console, a magnetic disk unit and a magnetic tape unit. A keystation consists of a conventionally laid-out keyboard, and a VDU or panel display. The keyed source data is transferred from the keystations to the magnetic disk via the miniprocessor. During this process the miniprocessor carries out checks on the source data and builds up control totals. It is also capable of providing guidance for each keystation operator in relation to the data set being input. All input data and guidance information appear on the VDU and thus the keystation operator has a clear picture of the stage reached and the errors detected.

Verification is by the re-keying of data in a similar way to punched cards; the system's more thorough checking of the source data enables verification to be omitted in some cases. A key-to-disk system allows erroneous or doubtful data to enter the system after it has been flagged for correction or completion in a later process. This philosophy promotes a more streamlined data entry system since there are less hold-ups and re-entry of data. After acceptance and verification the data is transferred from the magnetic disk to computer-compatible magnetic tape for subsequent input to a computer.

The keystations can be sited at dispersed locations up to 1000 ft from the miniprocessor. This facilitates the capture of data close to its sources such as factory machine shops. Some models of key-to disk systems can be fitted with communications equipment in order to transmit data to other key-to-disk systems or directly to the computer. In the latter case this makes the system effectively into a remote front-end terminal. Peripherals, such as card readers and line printers, can also be fitted, thus enhancing the system's power to that of a front-end processor.

The main disadvantage of key-to-disk systems is the possibility of breakdown of the miniprocessor causing all the keystations to be put out of action.

*Key-to-diskette systems*

A key-to-diskette system, e.g. IBM 3740, comprises a number of data stations each of which is equipped with a keyboard and a small VDU. The data stations are independent (stand-alone) in contrast to key-to-disk. The keyed data is displayed on the stations' VDU before being written to a diskette. A diskette is an eight-inch square plastic disk holding up to a quarter-million magnetic characters. The diskettes (floppy disks) are subsequently fed into a data convertor in order to transfer their data to computer-compatible magnetic tape.

The checking and guidance features of a key-to-diskette system are similar to those of key-to-disk but, because the data stations are stand-alone, they cannot act as front-end processors. The data convertor can however be fitted with communications equipment so as to make it on-line.

The floppy disks themselves are robust and light enough to be sent through the mail and passed from department to department. This enables the data stations to be sited in close proximity to the sources of data such as dispersed offices or branches.

### 10.4   Keying instructions

The documents from which data is punched or keyed come from a wide variety of sources. Their contents and layout are very diverse since they are connected with all aspects of commerce and industry. Among the range of documents we can expect to find all or some of the following:

> Pre-printed forms with handwritten entries, e.g. order forms.
> Lists printed from embossed plates, e.g. factory route sheets.
> Handwritten lists, e.g. computer programs.
> Printed documents, e.g. sales catalogues, tickets and labels.
> Sheets with ticked entries, e.g. transport surveys.

Whatever the source document, the system should be arranged so that only one document layout is used by an operator during one piece of work. If it is feasible, the source document is designed to facilitate the keying carried out from it but in practice this is often not possible because either the document already exists and cannot

152

be re-designed, or its layout is decided by its other purposes.

The data on a source document must be legible and identifiable; there is not much that can be done about illegibility except to create the data in some other way or take more care in writing. Identification can, however, be facilitated by having clear keying instructions allied to a specimen source document. For each type of source document a 'keying instructions' form should be filled in at the time when the related data input is first designed. Thereafter this form is held in the data preparation department for reference if there is doubt re the content and layout of source data. An example of a filled-in keying instructions form is shown in Figure 10.2, its entries relating to the specimen source document shown in Figure 10.1. The form in Figure 10.2 covers most requirements but can, of course, be modified to suit a particular user's needs; its contents are as under.

1   'Department' from which the documents originate.
2   'Queries to' — the name(s) of the person(s) who can help with queries regarding the content of a difficult document.
3   'Name' and 'Reference Number' (if any) of the source document.
4   'Name' of the data set or the card being handled; this is often the same as the document name.
5   'Reference' — see note on form.
6   Data item name — should correspond to those printed on the document, and/or card.
7   'Data item position' — the first and last positions of the data item's keying sequence or card columns.
8   'Picture of Field' — see notes on form.
9   'Justification' — this indicates the method for accommodating unfilled fields. Numeric fields are normally 'justified' right, i.e. with non-significant zeros at the left of the field; alphabetic fields are normally 'justified' left, with unfilled columns remaining blank to the right.
10  'Filling' — refers to the unfilled positions of the data item. A key-to disk system deals with justification and filling automatically according to its programmed format for the particular source data.
11  'Remarks' — usually unnecessary but any additional instructions can be inserted here.


*Timing of keying operations*

When estimating the time needed for the punching or keying of source data, an average figure of 10,000 key taps per hour may be used. This is an approximate figure modified in practice by the following factors.

1   The legibility of the source documents, especially if these are handwritten.

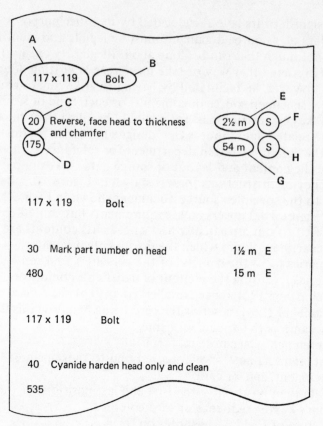

**Figure 10.1**  *Specimen source document*

2   The layout of the source document.
3   The number of queries raised, usually these are as a result of illegibility or missing entries, and the error procedures employed.
4   The type of data preparation or data entry equipment in use.

For each type of source data, it will be found from experience that a certain speed of keying can be maintained. In the planning stage the general rate of 10,000 key taps per hour, plus an allowance of one second per document, are satisfactory. If it is known that the source documents will be of poor quality, an additional time allowance should, of course, be made.

Thus if the daily keying load is 3000 documents from which an average of 60 characters each are keyed, the estimated time is

$$\frac{3000 \times 60}{10,000} \text{ hr} + 3000 \text{ sec} = 18 \text{ hr } 50 \text{ min}$$

This means that almost 19 machine hours keying and the same for verifying must be allowed. If this work load has to be completed

154

# KEYING INSTRUCTIONS FORM

| Department | | | | Queries to | | |
|---|---|---|---|---|---|---|
| Production control | | | | Mr. Underwood | | |

| Name of source document | | | | Ref. No. of source document | | |
|---|---|---|---|---|---|---|
| Operations master | | | | | | |

| Name of card | | | Electro No. | Colour stripe | Punch code |
|---|---|---|---|---|---|
| Operations master | | | 4-625 | Blue | |

| Ref. | Data item name | Data item positn | Picture | Justi-fication | Filling | Remarks |
|---|---|---|---|---|---|---|
| | Data type | 1 | 9 | – | – | = 6 |
| | Serial no. | 2 | 9 | – | – | Key sequentially within each part no. |
| A | Part No. | 3-9. | 999 A 999 | Right | Blanks | |
| B | Description | 10-19 | A ——— A | Left | Blanks | |
| C | Operation No. | 20-21 | 99 | Right | Zeros | |
| D | Machine No. | 22-24 | 999 | Right | Zeros | |
| E | Operation time | 25-27 | 99·9 | Right | Zeros | Key halves as 0·5 |
| F | Operation desgn. | 28 | 9 | – | – | Key E as 1, S as 2 |
| G | Set-up time | 29-31 | 9·99 | Right | Zeros | Hours and mins. |
| H | Set-up desgn. | 32 | 9 | –. | – | As F |
| C-H | 2nd operation | 33-45 | | | | As pos. 20-32 |
| C-H | 3rd. operation | 46-58 | | | | As pos. 20-32 |
| C-H | 4th operation | 59-71 | | | | As pos. 20-32 |
| | | 72-80 | B ——— B | – | – | |

**Notes:**

Ref.  Letter refers to data item shown on specimen source document

Picture  A = alphabetic field, 9 = numeric field,
X = alphanumeric field, B = blank

Figure 10.2  *Keying instructions form*

during 5 hours of the working day, then at least eight keying stations and eight operators will be needed.

The above example is obviously an over-simplification of a real situation. The\keying load is likely to be made up of many different batches of documents fluctuating in size from day to day. Nevertheless each significant batch should have its keying time estimated so that it can be fitted into the scheduling of routines (Section 11.3).

## 10.5   Non-keyed computer input

*Magnetic ink character recognition (MICR)*

This system has been developed as a means of providing printed figures that can be read by both humans and machines. By printing the figures in magnetic ink and in a stylized form (fount), this dual role becomes feasible. There are two MICR founts; the British and American standard is the E13B fount as is used by the banks; the E13B figures can be seen across the foot of bank cheques. In addition to the stylized numerals 0 to 9, there are four special symbols employed to signify the meanings of the data items.

The Continental standard MICR fount is known as CMC7 and differs from E13B in appearance. The range of CMC7 characters consists of the numerals 0 to 9, the alphabet, and the five special symbols. Each character is formed from seven vertical bars whose length and vertical position are arranged to form the character's appearance, and whose horizontal spacing enables the MICR reader to recognize the character. CMC7 comes in four sizes, E13B in one only.

The use of MICR documents for other purposes than as bank cheques is limited by the size restrictions imposed by MICR readers. The documents must be approximately six inches wide by three inches high and the stylized characters must be a quarter of an inch from the bottom edge. They are however also used on postal orders and luncheon vouchers, and usually employ CMC7 fount in this capacity. The original purpose of magnetic ink characters was to facilitate the high speed sorting of documents when returned for accounting procedures. It is now also possible to use MICR readers on-line and thereby feed their data directly into the computer. MICR sorter/readers operate at speeds of up to 2400 documents per minute, thus providing a fast means of computer input.

The magnetic ink is unaffected by being overwritten by ordinary ink, and the sorter/readers have a high degree of tolerance towards dirty and damaged documents; moreover, the nature of the ink makes forgery of documents difficult. However, owing to the restricted means of encoding MICR documents and their size and reading limitations, it is not likely that they will become widely used in business generally. The advances in optical character recognition make this a more suitable system for the majority of commercial and industrial applications.

## Optical character recognition (OCR)

This subject involves four types of optical readers, i.e. mark readers, character (document) readers, page readers, and reader sorters.

*Mark reading (OMR)* – the optical detection of small black marks on a document. The marks are roughly 6 mm wide, spaced 5mm vertically and from 16 to 24 may be made on one line.

The marks can be made in a number of ways.

1 Manually using a soft lead pencil or a reprographic pen.
2 By embossed plates, ordinary printing may also be done simultaneously from the same plate.
3 By a computer printer – printing hyphens.
4 Pre-printed when the documents are originally printed.

When marking the documents manually, conscious mistakes are eradicated by extending the erroneous mark downwards so as to form a small rectangle – the reader ignores this type of mark.

Documents are read at speeds of up to 100 documents per minute, but the actual speed attained depends upon the size of the documents being read. Most models accept a wide range of sizes up to about foolscap size. An important feature of mark reading is the flexibility of document layout. This results in a versatility of applications of the system. The marks do not have fixed values, meanings nor positions on documents; their interpretation is entirely by computer program.

Amongst the very wide range of existing applications of mark reading are the following:

1 *Customer order forms* Printed with rows or columns of commodity names with spaces for marking the order quantity alongside. The customer's account number, week number etc., may be printed by the computer as marks at the top of the form before sending it to the customer. This arrangement provides an almost foolproof ordering system because the minimum of manual marking is required. An elaboration of this principle is to produce the order document in the form of a diagram showing, for instance, an exploded view of an assembly or product. Each component thereon is arbitrarily numbered on the diagram and the required component is shown by putting a line through the appropriate number.
2 *Meter reading* These are generally pre-printed with the consumer number in the form of marks or optical characters.
3 *Survey questionnaire sheets* These must be very straightfor-

ward to mark if they are intended for use by the general public.

4   *Time sheets*   Used by the NCB and other large organizations.
5   *Examination answer sheets*   Suitable for questions which have a limited number of pre-stated answers.


*Document and page reading*

A document reader is limited to reading only a few lines of OCR characters per document; a page reader can read a full page of such characters. The main founts employed are:

OCR 'A' – American Standards Association.

OCR 'B' – European Computer Manufacturer's Association.
OCR 'A' has 66 different letters, numerals and symbols; OCR 'B' has no less than 113. Both founts have four standard character sizes.

Certain models of documents/page reader are capable of reading a limited range of hand-printed characters.

OCR documents lend themselves readily to certain applications and particularly those which can make use of turnround documents. These are printed in OCR fount by the computer, are sent out for external use and then returned for further processing by the computer. This system has tremendous advantages in that the indicative data such as the name and address, account code or part number, is absolutely accurate because it does not have to rely upon people's memory or being manually transcribed from other documents. Combined with mark reading and handprinting as the means of reinputting variable data, OCR documentation could become a standard procedure in industry and commerce.

Reader sorters read the data from the document and then move it to one of a number of stackers under computer control. Thus a limited amount of document sorting can be achieved.


*On-line input*

There are obvious advantages to be gained from the direct input of data to the computer, that is to say, without using any intermediate media or equipment other than a keyboard or visual display unit. This method involves the employment of terminals from which data is transmitted, and received after being processed on a time-sharing basis. Increasing use is being made of terminals as the means of prime data collection and this has the advantage of the early detection of errors of certain types. In particular, non-existent items are immediately detected because the computer cannot find the same on its files; thus the terminal operator can be prevented from continuing to key-in that particular transaction [10.29].

Where the terminal comprises a visual display unit, it is convenient and efficient to guide the operator by displaying alternatives from which the input data must be chosen. This would apply, for example, to an ordered item having a limited number of optional characteristics such as size or colour. In a similar way a user can be guided in obtaining information from files through a file interrogation procedure.

## Real-time input

The employment of terminal input-output is an absolute necessity with real-time business systems since by their very nature their data must be up-to-date. Real-time systems have come into prominence for airline reservation and banking purposes. The leaders in the field are undoubtedly the airlines; these accept airline seat enquiries and reservations on an interactive (dialogue) basis from VDU terminals in all parts of the world. Owing to the immediate response of such systems, error conditions and misunderstandings are largely obviated since the data is collected very close to its source, i.e. the prospective airline customer. Similarly the ability of all terminals to access and update the central flight reservation file avoids the possibility of over-booking and reduces the tendency to accept duplicated bookings.

Savings bank and building society systems are less conversational because the transactions are more stereotyped but nevertheless the input data (monetary transactions) immediately updates the central accounts file, consequently overdrawing, and fraud can be reduced if not eliminated entirely.

In addition to keyboard input, real-time and on-line systems can be arranged to accept input data of a more analog nature such as graphs or drawings by means of a light pen. Although this procedure is now well used for technical and scientific purposes, it is not likely to extend deeply into business applications for some considerable time, this is mainly owing to its expense and only the occasional need for it.

See [10.35 - 10.37] for further details of real-time systems.


## 10.6   Computer output

The prime method of outputting information from a computer is by printing. Other, less common, methods are visual display and micro-forms (COM).

*Printing*

Printers are of two main types: line printers and serial printers. The former are capable of printing a complete line of characters at a time, and their printing speed is independent of the number of characters actually printed per line. Line printers are subdivisable into drum printers and chain printers. A drum printer utilises a one-piece barrel embossed with the print characters. A chain printer consists of a belt of linked print slugs. Both of these arrangements rotate at high speed and the print is produced by a hammer striking a carbonised ribbon and stationery against the appropriate embossing or slug as it passes by. The practical difference between these two types of line printer is that with a chain printer the belt of slugs (chain) can be changed by the user so that different print founts are available.

Line printers operate at speeds between 300 and 2000 lines per minute, and so are suitable for preparing bulk printed output.

Serial printers print one character at a time by means of a print head moving across the stationery. Various methods exist for forming the printed character such as type slugs, wire matrices and ink jets. The speeds of serial printers are considerably lower than those of line printers, usually between 100 and 400 lines per minute.

Several other categories of printer are available, the most advanced of which is the laser/xerographic printer. This generates the print by means of an electronically controlled laser beam and this gives it a high degree of flexibility in the character set and fount shape and size. Its speed is around 13,000 lines per minute but this is tempered by the need to repeat the printing if copies are required.


*Visual display*

A wide range of visual display units (VDUs) are available, differing principally in the sizes of their displays. Normally something between 2000 and 4000 characters can be accommodated on a display. VDUs, although strictly output devices, are used more often in connection with the input of data. In this role they need to be capable of being edited so that the displayed data items can be corrected, deleted and others inserted.

The significant features of visual display that a systems analyst needs to bear in mind are:

1   *Intelligence* – the extent to which the VDU is capable of processing data independently of the computer.
2   *Editing* – the use of a movable cursor and a light pen to make changes to and select displayed data.
3   *Double brightness* – to differentiate between two lots of data, e.g. input data and data retrieved from storage, known as fore-

ground and background data.
4    *Split screen* — this allows two lots of data to be displayed and
edited separately, thus facilitating their comparison and amendment one from the other.

## Microforms (COM)

Computer output on microfilm (COM) is not used extensively for
computer output but nevertheless invites consideration when large
amounts of computer output are required. Its practicability depends
upon the output's subsequent usage. It is more suited to large volumes of archival information requiring only occasional interrogation.

Microforms are of two varieties — microfilm and microfiche. The
former is well known from the entertainment industry. A microfiche
is a rectangular sheet of plastic material, 6 x 4 inches, upon which
several hundred pages of information are optically recorded. Microfiche is more compact and easier to handle than microfilm. It is also
simpler to update because a microfiche can be extracted and replaced
by hand after alteration whereas a roll of microfilm has to be completely regenerated if one page needs altering.

The biggest difficulties in employing COM are the need for viewers
at the points of use, i.e. magnified displays, and the indexing problems
inherent to finding a particular page of information.

## Design of output documents and displays

The computer output is the end of the data processing chain and is
thus the 'product' of the data processing system. It is therefore imperative that the contents and layout are correct. In general, computer
output layouts are at present poor and consequently they create a
bad impression of data processing systems. To improve this situation,
the following points are worth bearing in mind:

*1    Clarity*    There must be no doubt as to the meaning of an item
of information on a printed document or a visual display. This means
that headings and annotations should be unambiguous, well positioned, and repeated on every page. If necessary, a heading is foot-
noted to amplify its meaning.

*2    Identification*    Each page must be entitled and consecutively
numbered. This obviates misidentification of information if a page
is separated from its partners.

*3    Layout*    The vertical and horizontal spaces between items should

be such as to provide ease of reading. This is especially so for management information as the reader may not have the opportunity of
becoming familiar with computer output. Care should be taken to
eliminate non-significant zeros and unnecessary operating systems
annotations.

A 'print layout chart' is a useful means of checking the appearance
and acceptability of a proposed document of display.

*4   Copies*   Printed output normally entails preparing several copies,
some of which may differ from others. The systems analyst must ensure that each copy is clear as regards its identification and intended
recipient. Coloured stationery with addressed headings facilitate these
aims.

*5   Paging/scrolling*   With visual display the output is in the form of
either separate pages or a continuous list of information items. The
former suits information falling into natural groups. The latter makes
comparison of information items easier because they can often be
positioned to appear on the same display. Scrolling also facilitates
the scanning of listed information.


## 10.7   Flowcharting

In systems design, the analyst becomes involved with the representation of data processing systems at various levels. At the highest level
a flow chart is used to illustrate the inter-connection of routines that
form the wholly or partially integrated system. An example of this is
shown in Figure 10.3; in this chart there is a box for each routine
together with inter-connecting lines. The solid lines are intended to
indicate the movement of information within the system, the dotted
lines indicate the causality between routines via the system's environment. Each routine should be clearly named, and it is helpful to code
each one for reference purposes.

The next lower level of flow charting involves a flow chart for
each routine. This shows the inter-connection of computer processing
runs and other operations (Figure 10.5) and utilizes a set of special
symbols. The simplified set shown in Figure 10.4 is extensive enough
for most purposes; these conform with British and International
standards [10.30].

The symbols are really intended to illustrate the concepts of the
routine rather than actual hardware devices. In simpler systems the
distinction between these two is somewhat blurred, but this is of no
consequence provided the intention is quite clear. The lines joining
the symbols represent the movement of data between processing
runs, and wherever possible these should be in one general direction,

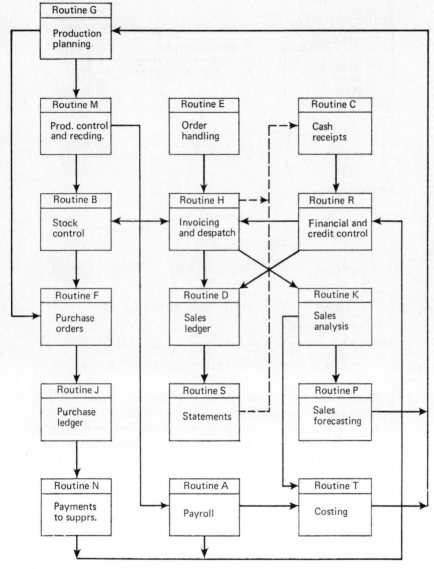

Solid lines represent movement of data between routines, broken lines represent data movement via outside agencies

**Figure 10.3** *Flowchart of overall system*

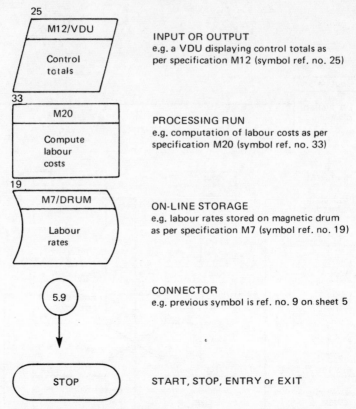

INPUT OR OUTPUT
e.g. a VDU displaying control totals as
per specification M12 (symbol ref. no. 25)

PROCESSING RUN
e.g. computation of labour costs as per
specification M20 (symbol ref. no. 33)

ON-LINE STORAGE
e.g. labour rates stored on magnetic drum
as per specification M7 (symbol ref. no. 19)

CONNECTOR
e.g. previous symbol is ref. no. 9 on sheet 5

START, STOP, ENTRY or EXIT

**Figure 10.4**   *Basic flowchart symbols (as per BS 4058:1973)*

usually from top to bottom, alternatively from left to right of the
flow chart.

Contained within each symbol is a code number by which it is
cross-referenced to its specification. As can be seen from Figure
10.5, it is convenient to prefix the symbol's code number with the
routine's code letter, and also to allocate the numbers in groups. This
allows for insertion of further numbers and helps to avoid confusion
between the files, runs, and documents. Where a file is common to
two or more routines, it always has the same code number — generally
allocated according to the routine that uses it most. Each symbol also
has a reference number near it in order to link it with other flowcharts
by means of the connector symbols.

## 10.8   Processing run specification

During the design of the data processing system, system flow charts
are prepared along the lines described in Section 10.7. These in themselves are of little use to the programmers, and it is necessary to explain in detail each processing run by preparing a specification of it.
This should be done in some depth so that the programmer is aware

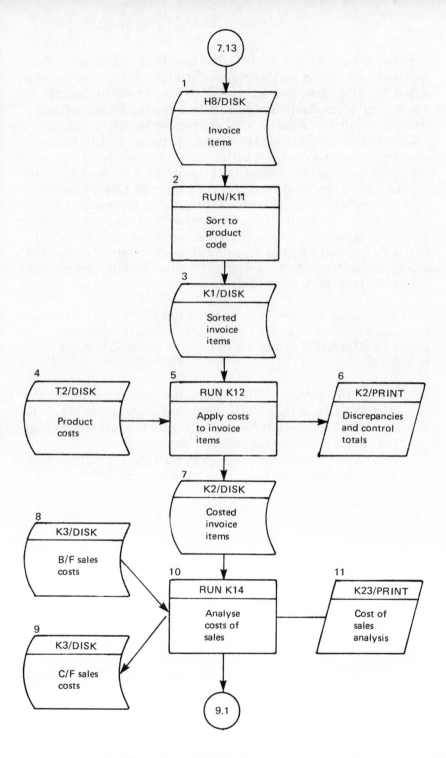

**Figure 10.5** *Flowchart of sales analysis routine*

of all the details. An example of a run specification is given in the appendix. Also, good documentation facilitates future amendments to the run. It is extremely aggravating to have to search through reams of program sheets, and to have to translate the instructions thereon, in order to discover what the run purports to do.

Quite apart from receiving a copy of the run specification, the programmer should discuss it with the analyst, and also ask questions about the run as he proceeds with the program. The worst situation is when a poor specification is presented to an unquestioning programmer; he is then apt to write the program in the way he vaguely believes to be the intention, only to find at the system testing stage that the program does not fulfil its real requirements. This sort of difficulty is avoided by not only providing a good specification but also by the systems analyst keeping in contact with the programmer during the program writing phase.

The four main features of a computer run are its input, processing, files, and output. These break down further into the following groups of information, all of which must be included in a run specification.

1   *General description*   This explains the purpose of the run in relation to the systems flow charts, and cross references it to the routine of which it forms a part. Any unusual terminology and special conditions should also be explained.

2   *Input layouts, punched cards*   For each class of card used as input, the undermentioned information is required:

Data item name
Card columns
Data item picture
Maximum and minimum values — per field in the card
Name of card
Class of card code

(grouped: Data item name, Card columns, Data item picture, Maximum and minimum values) per field in the card; all per class of card in the run

3   *Input layout, paper tape*   For each paper tape file used as input

Data item name
Date item picture
Symbols in the field
Maximum and minimum values — per field in the record
Record symbols
Tape labels and special symbols

(grouped: Data item name, Date item picture, Symbols in the field, Maximum and minimum values) per field in the record; all per tape file in the run

4   *Input layouts, marked documents*   For each layout of marked documents and for each marked row thereon:

Data item name
Data item picture
Maximum and minimum values of the data item
Value of each marking position (cell)
Relationship of cells to one another, e.g. are the marks additive?

5   *Input layouts, MICR and OCR documents*   The position of each

166

line of stylized characters together with a description of each (as per marked documents). The meaning of all inter-field symbols and special characters.

6   *Volumes of input*
Average and peak quantities of each class of card.
Average and maximum fields per spread card.
Average and peak numbers of each paper tape record.
Average and peak numbers of documents.

7   *Sequence of input*   Is the input in sequence, and if so what is the procedure if found to be out of sequence? Are there any serial numbers to be checked for continuity?

8   *Meanings of coded fields*   When it is necessary for the program to translate code numbers into a value or a descriptive form, these must be specified, for example:

| | |
|---|---|
| Factory Code | 1 = Coventry, 2 = Leeds, 3 = Bristol. |
| Discount Code | 1 = 20%, 2 = 15%, 3 = 10%, 4 = 5%. |
| Unit of measure | A = feet, B = square feet, C = pounds. |

9   *Stored data volumes*   This applies to items about which data is stored in core, such as tables, indexes and so on. The volumes of these items is helpful in allocating core accommodation.

10   *Processing*   A precise step-by-step explanation of the processes between the input of data and the output of results. These include all calculations, comparisons, tests, file amendments, etc., and where a calculation is at all complicated, the specification should include a worked example. With multiplications and divisions, the degree of accuracy and the method of rounding off are needed. Decision tables (Section 11.1) should be used to explain complex alternative conditions.

11   *Run flowchart*   As a general rule the systems analyst is well advised not to attempt to create a programming flowchart. It is both unnecessary and unhelpful since the programmer should be able to do this for himself, and in all probability better than the analyst. If a section of a run is difficult to explain in ordinary language, there is no harm in the analyst drawing a run flowchart as a means of conveying his meaning.

12   *Error checks*   Details of all feasibility checks (Section 5.3) and check digits (Section 5.4), and the procedures to be adopted if these fail; these may depend not only on the check itself but also on the position of the run when failure is detected.

13   *Control totals*   All control batch totals and hash totals are to be accumulated, with an explanation of any unusual hash totals. Control totals include both input totals for checking, and output totals for printing (Section 11.4). Action to be taken on finding a discrepancy in an input control total.

14   *File layouts and organization*   For each existing file that is to be accessed during the run, and for each new file created by the

run, the details needed are:
>Number of records — now and in future.
>Contents and layout of each type of record.
>Mode of storage and sequence.
>Key(s) within records.
>Magnetic tape block sizes.
>Method of addressing direct access buckets.
>Organization and sizes of cylinders, buckets and indexes.
>Allowance for and organization of overflow.

When the run involves the creation of a new file, it is important that its characteristics are agreed by both the analyst and the programmer, and also that they follow accepted file standards within the data processing department. If a database is in use, the database administrator decides this point.

15  *Security arrangements*   At what points of the run and under what circumstances dumping is to be performed. What files are to be duplicated (if any), and a description of any other special arrangements for the protection of files or other data. This again involves the database administrator.

16  *Output layouts*   Output in the form of punched cards and paper tape is specifiable in the same way as for input.

Printed output is most satisfactorily specified by the use of print layout forms. These are to scale and so facilitate the design of pre-printed stationery. When the output is to be on blank stationery, the print layout ought to include the column headings to be printed at the top of each sheet. Other features of printed output are:
>The number of copies to be printed at a time.
>Sheet numbering method.
>Vertical spacing between lines of print.
>Detailed rules for throwing to next sheet so as not to
>divorce totals from their itemized sections.

17  *Sizes of totals*   An indication of the maximum of totals — both intermediate and final, obviates the loss of the most significant digits of a total. This can occur owing to either overflow from the core locations assigned to a total, or failure to allocate sufficient printing positions for it.

## 10.9   References and further reading

*Form design*

10.1   *The design of forms*, HMSO (1962).
10.2   'Making top form', *Data Systems* (May 1971).

*Systems design*

10.3    CLIFTON and LUCEY, *Accounting and computer systems*, Business Books (1973).
10.4    MATTHEWS, *Design of the management information system*, Chapters 2-5, Auerbach (1971).
10.5    CLIFTON, *Data processing systems design*, Business Books (1971).


*Data input*

10.6    *Alphanumeric punching codes for data processing cards*, BS 3174 (1959).
10.7    *Data processing codes for punched tape*, BS 3480 (1962).
10.8    SMYTHE, *Guide to computer preparation*, Business Publications (1970).
10.9    RUBIN (Editor), *Handbook of data processing management*, Volume 4, *Advanced technology — input and output*, Auerbach (1970).
10.10   'Data entry', *Data Systems* (May 1976).
10.11   'Data preparation', *Data Processing* (July-August 1970).
10.12   'Key-to-disk', *ibid.* (September-October 1974).
10.13   'The end of paper tape', *Data Systems* (April 1971).
10.14   'Magnetic tape encoding', *ibid.* (April 1971).
10.15   'Key-to-tape', *Business Automation* (June 1970).
10.16   Data preparation supplement, *Data Systems* (March-April 1974).
10.17   'Source data collection', *ibid.* (February 1973).


*Character and mark recognition*

10.18   *Character recognition*, British Computer Society (1971).
10.19   'OMR for sales order processing', *Data Processing* (January-February 1972).
10.20   'OCR — benefits and pitfalls', *Computer Journal* (May 1969).
10.21   'OCR — a practical case', *Data Processing* (July-August 1971).
10.22   'Super scale OCR', *ibid.* (September 1970).
10.23   Character recognition supplement', *ibid.* (September-October 1973).
10.24   'Mark reading', *ibid.* (July-August 1969).
10.25   'Decentralised control of order processing', *ibid.* (September-October 1973).
10.26   'Optical character recognition', *Computer Automation* (November 1970).

10.27 'Baker finds crumb of comfort in OMR', *Data Systems* (June 1973).
19.28 Character recognition', *Data Processing* (May-June 1970).
10.29 'The utilisation of graphic display units as the main form of computer input', *Computer Journal* (May 1969).


*Flow charting*

10.30 *DP flowchart symbols,* British Standard, BS4058(1973).
10.31 CLIFTON, *Data processing systems design,* Business Books (1971).
10.32 'Flowcharts for all', *Data Systems* (September 1970).
10.33 LEHNER, *Flowcharting: An introductory text and workbook,* Auerbach (1972).
10.34 WAYNE, *Flowcharting concepts and DP techniques*, Canfield (1973).


*Real-time*

10.35 MARTIN, *Design of man-computer dialogues,* Prentice-Hall (1973).
10.36 MARTIN, *Design of real-time computer systems,* Prentice-Hall (1969).
10.37 MARTIN, *Systems analysis for data transmission,* Prentice-Hall (1972).

# Chapter Eleven
# *Design of data processing systems-2*

## 11.1 Decision tables

When preparing a computer processing run specification it is some-times difficult to describe verbally certain requirements of the run. As mentioned in Section 10.8, this problem is alleviated by the use of a detailed run flow chart, and quite often this is the most satisfactory solution. This is particularly true when the run involves 'looping', i.e. repetition of the same series of steps a number of times. Another problem is when the run includes a number of 'decisions' which result in branching of the steps. A few such decisions following one another give rise to a large number of branches resulting in a complicated and extensive flow chart or verbal description.

The difficulty of dealing with this abundance of branches is alle-viated considerably by the employment of 'decision tables'. A deci-sion table is a method of specifying the branching rules of a process-ing run, and is used either instead of or in addition to a detailed flow chart. There is no logical reason against specifying the complete run in the form of a decision table. Quite often though this would intro-duce unnecessary work as the descriptive form is adequate, and often clearer to understand.

A decision table is set down in the form of a table of figures and words subdivided into four parts (separated by double lines), as shown in Figure 11.1. Of the four parts, the upper two, 'condition stub' and 'condition entries', describe the conditions that are to be tested. The two lower parts describe the actions to be taken depen-dent upon the outcome of the tests. A 'rule' is in a single vertical column of the table, and consists of a set of outcomes of condition tests together with the associated actions. The condition entries obey

'AND' logic; that is to say if there are two or more condition entries in a rule's column, all the conditions must be satisfied for the rule to apply. Similarly 'AND' logic is applicable to the action entries of a rule, and in some cases the sequence of the actions is also important.

*Types of decision table*

The entries in a decision table may be in either limited or extended form; this applies to both conditions and actions. A limited entry is where the stub completely specifies the condition or action; the condition entry then consists of 'yes', 'no'. or a dash (meaning irrelevant); the action entry consists of a cross if applicable, or a blank otherwise. An extended entry is where the specification is only partially made by the stub and is completed by the entry.

There are four arrangements that are used:
1    Limited conditions and limited actions.
2    Limited conditions and extended actions.
3    Extended conditions and limited actions.
4    Extended conditions and extended actions.

Examples of these four arrangements, relating to the flow chart in Figure 11.2 and to the descriptive specification that follows below, are shown in Figure 11.3.

Suppose we are applying a decision table to a production control application in which it is necessary to decide the interval times to be allowed between manufacturing operations. In ordinary language the rules to be applied could be stated thus:

'If the operation is the first for the component being made, the interval is zero. Otherwise if the machine for the operation is the same as that for the previous operation and the machine number is 99 or less, the interval is again zero. When the machine for the operation differs from that for the previous operation and the machine number is 99, the interval is half a day; if the machine number is less than 99, the interval is one day. All operations on machines numbered 100 upwards are allowed an interval of four days.'

Although this statement is accurate, it is not easy to follow. The requirements are more easily understood from Figure 11.2 or when shown as one of the four alternative arrangements in Figure 11.3 [11.1-11.10].

## 11.2  Computer run timing

The reasons for estimating the computer run times when designing a data processing system are threefold;
1    The estimated time serves as one basis for comparing the various

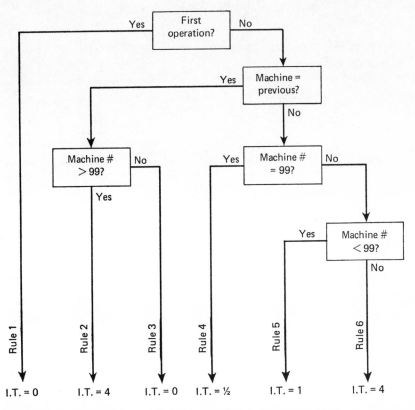

Figure 11.1

**Figure 11.2** *Flowchart of decision tables in Figure 11.3*

173

(a) Limited condition/limited action decision table

| Rule | 1 | 2 | 3 | 4 | 5 | 6 |
|---|---|---|---|---|---|---|
| First operation ? | Y | N | N | N | N | N |
| Machine = previous? | – | Y | Y | N | N | N |
| Machine # = 99 ? | – | – | – | Y | – | – |
| Machine # < 99 ? | – | – | – | – | Y | – |
| Machine # > 99 ? | – | Y | N | – | – | Y |
| Interval time = 0 | X | | X | | | |
| Interval time = ½ | | | | X | | |
| Interval time = 1 | | | | | X | |
| Interval time = 4 | | X | | | | X |

(b) Limited condition/extended action decision table

| Rule | 1 | 2 | 3 | 4 | 5 | 6 |
|---|---|---|---|---|---|---|
| First operation ? | Y | N | N | N | N | N |
| Machine = previous ? | – | Y | Y | N | N | N |
| Machine # = 99 ? | – | – | – | Y | – | – |
| Machine # < 99 ? | – | – | – | – | Y | – |
| Machine # > 99 ? | – | Y | N | – | – | Y |
| Interval time = | 0 | 4 | 0 | ½ | 1 | 4 |

(c) Extended condition/limited action decision table

| Rule | 1 | 2 | 3 | 4 | 5 | 6 |
|---|---|---|---|---|---|---|
| Operation | First | Not first | Not first | Not first | Not first | Not first |
| Machine | – | As previous | As previous | Not as previous | Not as previous | Not as previous |
| Machine # | – | > 99 | ≤ 99 | = 99 | < 99 | > 99 |
| Interval time = 0 | X | | X | | | |
| Interval time = ½ | | | | X | | |
| Interval time = 1 | | | | | X | |
| Interval time = 4 | | X | | | | X |

(d) Extended condition/extended action decision table

| Rule | 1 | 2 | 3 | 4 | 5 | 6 |
|---|---|---|---|---|---|---|
| Operation | First | Not first | Not first | Not first | Not first | Not first |
| Machine | – | As previous | As previous | Not as previous | Not as previous | Not as previous |
| Machine # | – | > 99 | ≤ 99 | = 99 | < 99 | > 99 |
| Interval time = | 0 | 4 | 0 | ½ | 1 | 4 |

**Figure 11.3** *Types of decision table*

methods of achieving the desired results.

2    It helps the systems analyst to decide what computer configuration is needed in order to carry out all the applications within their processing cycles.

3    The scheduling of computer runs becomes possible, thus giving an estimate of the over-all work load on the computer and the data processing operating staff.

In relation to the first reason, there are invariably several methods for achieving the required outputs. Provided there are no overriding objections, they should all be time-estimated as a basis of comparison.

*Factors involved in run timing*

The estimation of computer run times is a fairly complex subject involving a large number of factors. They can be summarized into the following main factors:

1    The volume of data to be processed in the run during input, output and intermediate stages.

2    The computer configuration, especially in relation to the speeds of its peripherals.

3    In the case of a small computer, the amount of overlap of processor and peripheral operations, the degree of multi-programming that is possible, and the efficiency of the operating system (if any).

4    The types of files, their size, access times, transfer rates and modes of storage and access.

5    The efficiency of the program — this obviously includes the programmer's skill but is far more affected by the programming language and the efficiency of the compiler.

We must bear in mind that at the system design stage the analyst cannot expect to obtain absolutely accurate timings. An estimate within 10 per cent is perfectly satisfactory since it is unwise to load the computer to its limit anyway.

For a given computer run, all or some of the following times contribute to the overall run time. Their relationships are, however, somewhat intricate and dependent upon the particular model and configuration of computer being used.

1    *Peripheral times*    These are mostly the times to read cards, documents and paper tape, the printing time, and any other input/output operations.

2    *Magnetic tape*    Reading, writing, and sorting times.

3    *Direct access storage times* — access to records, and their transfer to and from core store (Section 7.6).

4    *Processing time*    Including calculations, index searching, data organization, internal sorting, and internal housekeeping.

175

5    *Setting-up time*    A wide variety of factors contribute to this, and their significance depends upon the mode of operation of the computer. With multi-programming controlled by an operating system, the setting up time effectively disappears, whereas with a small batch processing computer doing a large number of different runs, setting-up can take a significant proportion of the total time. Included in setting-up time are:

a    The reading into core of the program from magnetic tapes, disks, cards or paper tape.

b    The loading of peripherals with cards, paper tape and stationery.

c    The loading of magnetic tape reels and removable disks.

d    The keying of messages on the console typewriter, and setting switches.

The time to allow for setting-up is from zero to five minutes per run, depending on the above factors.

6    *Running allowance*    This covers all the extra time consumed by eventualities such as control checks, mis-reads, error prints, tape rewinds, dumping, and so on. The allowance again depends upon the mode of operation and the complexity of the run; up to twenty per cent of the estimated time should be added.

*Over-all timing method*    This method involves the estimation of the time for a run by considering the total volume of data to be handled by each of the units of the computer. In this task the manufacturer's timing manual is invaluable and from it each unit's time can be assessed. The degree of overlapping of activities (simultaneity) of the configuration must also be established, and when this occurs it is generally the longest of the overlapping times that matters. For instance, if three activities such as card reading, magnetic tape reading, and magnetic tape writing, occur concurrently during a run, the extent to which these overlap must be found from the manufacturer's literature. Very probably it would be the card-reading time that predominates, the tape operations being absorbed into the time gaps between card operations.

Where the computer is capable of multiprogramming operation, it is essential to balance the programs being run concurrently. This is catered for automatically if an advanced operating system is employed but otherwise it devolves upon the systems analyst to attempt to balance the jobs. This means that as far as is possible peripheral-bound jobs should be run along with processor-bound jobs so that the computer's full potential is being utilized all the time. This arrangement may, of course, conflict to some extent with the run scheduling (Section 11.3) as dictated by other factors such as source data availability and preparation, post computer jobs, and output deadlines. This principle of run balancing still holds true however and should be taken into consideration when preparing the run scheduling.

Having established the computer's ability to overlap operations, the nominal time for each of the computer's peripherals and other units has to be calculated. This is done quite simply but certain factors should be taken into consideration.

*Input timing factors* With punchard cards, it is not only the number of cards but also to some extent the amount of data in them that is of interest. Spread cards containing a large number of items need more data distribution, and this can cause the card reader to fall below full speed quite apart from the processing of the items. This is not to say that spread cards give a net slower input; the reduction in speed is minimal whereas the increase in card contents is usually considerable. A typical example is a speed reduction to say 80 per cent, and an increase in contents to five items per card, giving a net increase in data input speed of four times.

*Magnetic tape timing factors*
1    The effective transfer rate taking into account the block length and interblock gap length. The significance of these can be seen from Figure 11.4, which shows the times taken to read 50,000 records of 100 characters each from a 41,700 characters per second tape deck.

| Block length, characters | Read time, min, with inter-block gap of | |
|---|---|---|
| | 0·56 inch | 0·75 inch |
| 100 | 11·8 | 13·5 |
| 500 | 4·0 | 4·3 |
| 1,000 | 3·0 | 3·2 |
| 4,000 | 2·2 | 2·3 |

**Figure 11.4**    *Magnetic tape read times*

2    Whether each reel of tape must be rewound before another is brought into use. This situation applies when only one tape deck is available for a file. If two decks are available, the file's tape reels are 'queued' on alternate decks thereby saving rewind time because one reel can be rewound while the other is being read or written. The operating system takes care of this situation.

3    The sorting time taken by the magnetic tape is important, and in some routines this dominates the overall time. Tape sorting timing-tables are available from the manufacturers, and these take into account the factors below.

Transfer rate (read/write speed).

Number of decks used in the sort.
Amount of core store available as buffer areas.
Input/output channels available.
Transfer direction(s) — can the tape be read backwards?
Block length and inter-block gap length.
Number and size of records to be sorted.

*Direct access storage device timing factors*    This is covered in Section
7.6 from the logical aspect. The hardware aspect factors are the access
times and the transfer rates. The former may include the times taken
to position read/write heads, verification of head positioning, and
rotational delay of the storage media. The transfer rates tend to vary
even on the one device, depending on the location of the required
data. Great care must be taken when comparing these devices that all
these factors are understood and taken into consideration.

*Output timing factors*    The factors entering into the timing of line
printers are:

1    The number of lines to be printed.
2    The spacing between the lines and the amount of skipping
     (throwing) per document. Skipping may occur over the station-
     ery perforations and also in the body of the document.
3    The skip rate of the line printer.
4    The range of printed characters needed; this applies to certain
     models of drum printers.
5    The width of the print bank; this in itself does not affect the
     printing speed, but with a wide print bank it may be possible to
     print two or more documents at the same time (spread printing).

Serial printers involve the following factors:

a    The number of characters printed per line.
b    The line advance and carriage return time.
c    The bi-directional printing capability of the printer; this reduces
     the carriage return time.
d    The amount of skipping and the skip rate of the printer.
e    Dual carriage fitment for printing two documents concurrently.

*Example of run timing*    The principles behind run timing are best
seen from a simple example; this is shown in Figure 11.5 and consists
of a routine of three computer runs assumed to be carried out
separately on a fairly small computer without an operating system.

Run 1    Read 20,000 spread cards, write the contents to magnetic tape A1, and
         at the same time print details from certain cards amounting to 3000 lines
         of print. The magnetic tape is written so as to hold 50,000 records of 40
         characters in 400 character blocks, i.e. each spread card creates an
         average of 2.5 records, duplication of certain data occurring.

Run 2    Sort the 50,000 records to produce tape A2.

178

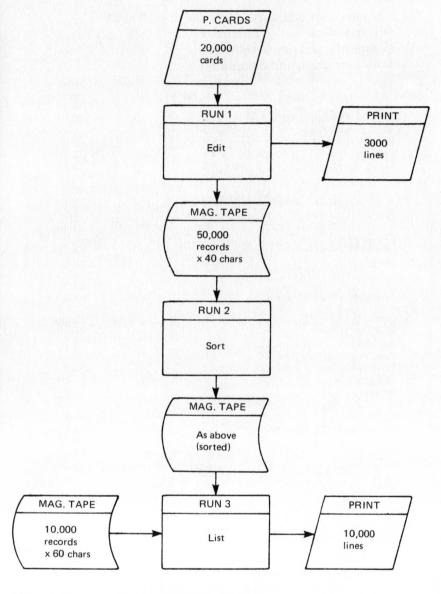

**Figure 11.5** *Flowchart of run-timing example*

*Run 3*  Read the sorted tape A2 and the pre-sorted tape B, the latter holds
10,000 records of 60 characters in 900 character blocks. Summarize
the records in tape A2, and amalgamate with the records from tape B so
as to print 10,000 lines.

The computer configuration is:

| | |
|---|---|
| One card reader | – 900 cards per minute. |
| One line printer | – 600 lines per minute. |
| Four magnetic tape decks | – 20,800 characters per second. |
| 32,000 characters of core storage. | |

The computer's simultaneity covers the operation of:
   One tape deck and one peripheral.
   Computing and one peripheral.
   Four tape decks and computing.

| | Nominal time, min | Notes | Allowed time, min |
|---|---|---|---|
| *Run 1* | | | |
| Read 20,000 cards at 900 c.p.m. | 22.2 | | 22.2 |
| Write 50,000 records | 3.1 | a | — |
| Print 3000 lines at 600 l.p.m. | 5.0 | | 5.0 |
| Compute | 4.5 | b | — |
| Set-up | 4.0 | | 4.0 |
| | | | 31.2 |
| 10 per cent running allowance | | c | 3.1 |
| Rounded total | | | 35 min |
| *Run 2* | | | |
| Sort 50,000 records | 58.0 | d | 58.0 |
| Set-up | 3.0 | | 3.0 |
| | | | 61.0 |
| 5 per cent running allowance | | e | 3.0 |
| Rounded total | | | 64 min |
| *Run 3* | | | |
| Read 50,000 records (tape A2) | 3.1 | | 3.1 |
| Read 10,000 records (tape B) | 0.7 | f | — |
| Print 10,000 lines at 600 l.p.m. | 16.7 | | 16.7 |
| Compute | 3.8 | g | — |
| Set-up | 4.0 | | 4.0 |
| | | | 23.8 |
| 5 per cent running allowance | | h | 1.2 |
| Rounded total | | | 25 min |
| Total time for the routine | | | 124 min |

*Notes*
a  –  overlapped with card reading.
b  –  overlapped with printing.
c  –  for card handling and stationery adjustments.
d  –  time derived from manufacturer's manual.
e  –  for tape handling.
f  –  overlapped with tape A2 reading.
g  –  overlapped with printing.
h  –  for stationery adjustments.

When a multiprogramming system and an operating system are in use, it is difficult to estimate run times. Under optimum conditions the set-up times largely disappear as setting up occurs when the peripherals are not in use. Similarly, the effective simultaneity of the computer is greater because the peripheral-bound jobs are balanced and the input/output data is spooled. Nevertheless, it is safe to estimate the job times in the manner described above as they will not be exceeded in practice.

## 11.3    Run scheduling

The scheduling of computer runs is done in outline during the system design stage in order to ensure that all the work can be fitted into the available computer time. Later, during the implementation stage, the runs will be rescheduled so as to form a precise time-table of computer operations. Before attempting to create a schedule, each routine will need to have been split into a number of computer runs, and each run to have had its time estimated.

As stated in the previous section, if the computer to be employed is capable of operating in multiprogramming mode, several processing runs can be carried out simultaneously. The details that follow do not, in fact, allow for multiprogramming but assume that only one job at a time can be run on the computer. To run schedule a multiprogramming computer, it is first necessary to find the information appertaining to points 1, 2 and 4-6 below and then to balance groups of jobs for concurrent scheduling. This process is rather more complicated and is not worth planning in great detail at this stage of systems design; in reality job scheduling will become a tactical exercise later for the chief operator and the operating system. In addition to the processing runs, the times taken for data preparation need estimating, and the times of availability of source documents determined. Some of the routines have a deadline time by which the output results must be ready; failure to meet these results in delays in dispatching, production, payment of staff, and so on. These deadline routines are given highest priority in the scheduling procedure.

For each routine in the programme of work the following information is required before scheduling is done:

1    At what time are the source documents available, and are they in one batch or do they appear in a steady stream?
2    How long does data preparation take?
3    How long does each computer run take?
4    What are the deadline times for final and intermediate results?
5    Is the output extensive enough to justify splitting it into batches for dispatching – this ties in with the viability of run splitting (Section 10.1).
6    What extra work must be done after the computer output is ready and how long will it take? This work includes tasks such as the decollating, bursting and trimming of stationery, the checking and editing of results, the enveloping and dispatching or delivery of the output.

With the above information to hand, the high-priority routines are fitted into the schedule, followed in turn by those of decreasing priority. The high-priority routines themselves are best dealt with in order of spare time, those with the least spare time being scheduled first. In this context spare time is the difference between deadline time and source document availability time less the times for data preparation, computer running and post-computer work.

When a routine has spare time available, it can be shifted about within its spare time in order to facilitate the scheduling of other routines. It is preferable not to separate the runs forming a routine but to carry them out in close succession; if separation is unavoidable, it is applied to the lower priority routines first. When scheduling a routine within its spare time, the analyst is torn between making it as early as possible and thereby risking the disruption of other routines if the source documents are late, and leaving it as late as possible, risking delayed output results if trouble occurs. A way round this dilemma is to adopt, if possible, a compromise schedule for the routine. This is decided by balancing the probability of documents being late against the probability of computer failure. Since it is unlikely that either of these probabilities are known precisely, a reasonable plan is to split the spare time fifty-fifty before and after the routine. Once a routine has been scheduled, it may well be necessary to move it within its spare time so that lower priority routines can also be accommodated. An example of this is seen in Figure 11.6, in which the Wednesday payroll computing has been moved forward so that production planning can be fitted in before its deadline.

*Computer run scheduling chart*

This is shown in Figure 11.6 and is based on the computer's switched-on time. This depends upon the number of shifts that will be worked; a fairly safe assumption is initially two shifts amounting to 12 hr per day for five or six days of the week. Out of this time a number of hours is lost to program testing, scheduled maintenance, computer breakdowns (sometimes known as 'unscheduled maintenance'), re-runs due to breakdown, program failure, erroneous input data and operating mistakes. The manufacturer will specify the time needed for scheduled maintenance and might even venture a suggested allowance for breakdown. Program testing tends to decrease as more applications become tested and implemented on to the computer, so that to some degree these two times tend to balance out. It is difficult to allow at all accurately for re-runs but fortunately they tend to diminish with time; initially an allowance of 25 per cent of productive time is not untoward.

The routines are dealt with in order of priority, and it is convenient to work from the top of the chart downwards. A certain amount of manoeuvring is needed to fit in the computing and data preparation – this is facilitated by using a chart with movable indicators. Under each routine the accumulative 'bookings' of both the computer

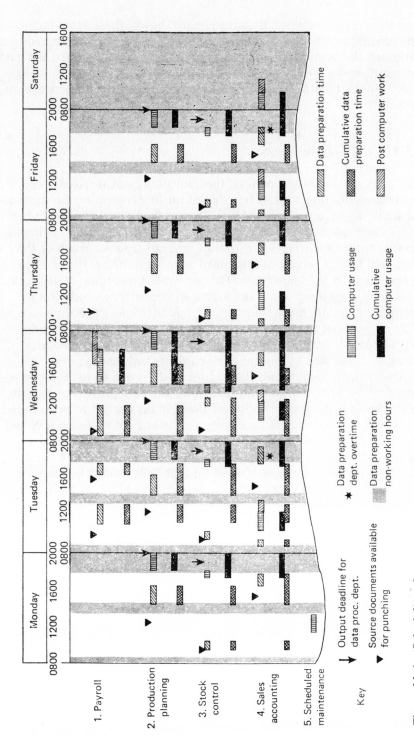

**Figure 11.6** *Scheduling chart*

183

and the data preparation department are shown; these again facilitate the scheduling of the remaining routines. It must be remembered that although the computer works from say, 0800 to 2000 hours, the data preparation department normally works from about 0803 to 1200, and 1400 to 1730 hours; outside these times they would be working overtime.

## 11.4 Audit requirements

The responsibilities and aims of the auditor are not changed when the routine accounting work is transferred on to a computer. His methods of achieving his aims and fulfilling his responsibilities do change however. The tendency is for his work to involve less checking of figures, leaving more time to investigate the causes of trouble. Much of the data that he has become accustomed to using during auditing is no longer there in the same form. The permanent recording of all transactions no longer occurs as the nature of computer recording is to make interim data more transient. The auditor can, nevertheless, obtain details of the system, and although the 'audit trail' in some ways becomes less distinct, this is counterbalanced by the auditor being less confused by masses of documents and figures.

The systems analyst and the auditor both have a vested interest in ensuring the accuracy and the validity of the data processing system. It is therefore the responsibility of the systems analyst to provide the control totals and checks necessary for the auditor to do his job, and the duty of the auditor to adjust his approach to his work so as to dovetail into data processing methods. In order that the auditor can do this with confidence, it is essential that he is trained in computer techniques. Although he cannot be expected to know as much as the analyst or programmer about the detail of a particular system or program, an understanding of the methods therein is vitally necessary. This means that the auditor must know the meaning of data processing terminology and the purposes of hardware and software.

*System security controls*

The perpetration of fraud in a data processing system is not impossible but is extremely difficult for one person if comprehensive controls are built into the system. A successful fraud would require the connivance of several persons together with a considerable amount of skilled effort on their part. This would be far more so than in a semi-manual system where there is far more opportunity to make unauthorized and undetected alterations to documents during the course of their life.

A more pertinent feature of systems control is the detection of accidental errors. These may be due to a number of different causes, such as:

Omission of input data.

Obsolete input data.

Incorrect master files — incorrectly updated previously or wrong file used.

Undetected program errors — especially after program amendments.

Malfunctions of hardware — especially misreading and misprinting.

Out-of-sequence transactions or master files.

Misunderstanding or careless use of operating instructions.

What can the systems analyst do to reduce the chance of accidental errors and fraud in a data processing system?

1   A count of all documents sent for keying, and as far as is reasonable, counts of the items thereon. The documents should be batched and accompanied by a batch slip on which are written the counts.

2   Batched totals of quantities and values on documents also included on batch slips. This may involve too much manual work if every field is totalled, the most significant only are therefore included. If no fields of the document are truly accumulative, a 'hash' total is formed by adding other fields such as code numbers.

3   The above-mentioned counts and totals are checked after the data has been punched or entered. Most data entry/preparation equipment creates control counts and totals automatically during keying.

4   At the end of each batch in the computer input run, the counts and totals are printed; alternatively they may be stored and printed as a table after reading all the batches. As an additional security measure the control counts and totals themselves may also be inputted and checked automatically against the figures accumulated by the computer. The details of any discrepancy are then not only printed but, if desired, the computer is programmed to 'lock' so that special operator action has to be carried out before the run can proceed. There is then less chance of ignoring a discrepancy — deliberately or unintentionally.

5   After each run in a routine, the control counts and totals are again printed including any new figures created as a result of the processing. The latter are especially important if they relate to data that is used in the subsequent runs.

6    Amendments to or the updating of a master file should be accompanied by a simple analysis (control account) of the master file before and after the run. This includes the numbers of records and any suitable data item counts and totals. The analyses are dated and retained as a visible history of the file; they can also be held in the file itself to cover a period of a year or so. An example of a control account:

| | | |
|---|---|---|
| B/F FILE | 12,750 records | £13,895 opening balance |
| INSERTIONS | 393 records | £4,250 |
| DELETIONS | 258 records | £2,631 |
| AMENDMENTS | 606 records | £1,580 added |
| C/F FILE | 12,885 records | £17,094 closing balance |

7    An operations log is prepared by the operating system or manually. This is preserved for reference purposes so that the auditor can query the reasons for re-runs and assure himself that no un-authorized computing has occurred.

8    All computer printed output must be clearly identifiable in terms of its exact particularity. This entails the precise labelling of every sheet of print with its appropriate captions, date, page number and, in certain instances, line numbers also. Page and line numbers reduce the chance of loss or fraud.

9    Make an audit package available, thus utilising the computer to check the system. Audit packages contain the following features:
*a*    Extraction of file data by variable parameters or sampling.
*b*    Test data for the spot checking of computer runs.
*c*    Summarisation of data items from all or selected records.
*d*    Calculation accuracy checks.
*e*    File completeness checks to detect loss or duplication of records.
*f*    Open-endedness to allow the auditor to modify or extend the package to meet his precise requirements.

10    The security arrangements for on-line and real-time systems are more exacting and involve stringent control over the use of terminals. Each terminal operator must be identified to and accepted by the computer before being permitted to have access to files. Additionally real-time involves complex procedures for dealing with breakdowns so that no messages are lost or duplicated, and so that audit require-ments are satisfied. This is too large a subject to cover herein. The reader is advised to consult Chapter 35 of [7.6] for full details.

11    Arrange that only authorised staff are allowed to operate the computer or terminals. Operation of the mainframe is normally pro-hibited to programmers. Terminal operators need to identify them-

selves absolutely; thereafter the computer controls the file data to which they have access.

12    Document all routines, runs and programs completely and keep master copies under lock and key. This prevents unauthorised amendments to programs.


## 11.5    Determination of computer configuration

The determination of run times and the subsequent scheduling of the computer's work load are based upon the capabilities of a particular configuration of computer. This may be a hypothetical configuration when the timing and scheduling are first planned, and these tasks should be repeated for several such configurations. The characteristics of these hypothetical computers should be based on those available with real computers so that when the most suitable has been decided upon, similar actual configurations can be given consideration. This means, of course, that the systems analyst must be fully aware of the range of models and peripherals that are on the market.

The power of a business computer is largely determined by two main factors:
1    Its file storage capacity.
2    The speed of its peripherals and storage devices.
In deciding at what levels these should be chosen, the analyst must look to the future as well as the present. As estimate of future volumes of computer work, based upon a forecast of the company's future activities, may tip the scales in favour of a particular model and configuration. In a similar way it may be discovered that the times taken for most of the runs hinge upon the speed of one peripheral; this is an obvious case for considering more or faster peripherals of this type.


*Main (core) store*

Main store, like education, cannot be seen overtly but its deficiency makes life much more difficult. Although there is no formulative relationship between the amount of core store and the general size of the computer, the two tend to increase together. When estimating the required amount of core store, the systems analyst must ensure that there is sufficient to accommodate not only the programs but also the other software. This includes operating systems, executive, control and supervisory programs, application packages, sorting and merging programs. Allowance should also be made for buffer areas, indexes, tables and other storage areas.

*Magnetic tape*

A glance at the manufacturers' manuals shows that the time taken for sorting records held on magnetic tape is related to both the numbers of tape decks used and the transfer rate of the tape. The decrease in time taken is most marked between three and four decks, the extra deck reducing the time by roughly 25 per cent. With additional decks this decrease is less marked, so that unless a large amount of sorting has to be done, four decks are sufficient. Sorting speed is more or less proportional to the tape transfer rate, so the problem of sorting is one of balancing the number of decks against the tape transfer rate chosen.

*Disks and other direct access storage devices*

The three main criteria to be applied to the selection of direct access storage devices are storage capacity, access time and transfer rate. Storage capacity is closely connected with the exchangeability of the storage media, so that it is the on-line storage requirement that is the determining factor. Dealing with the latter two criteria, their degree of importance depends upon the modes of file access and storage, and upon the files' activities. The access time and transfer rate must be consistent with the requirements of the system, but generally, provided the files are properly organized, they do not significantly affect computer run times of most conventional business applications.

Returning to storage capacity, bearing in mind the cost of disk packs, their storage capacity should not be wasted. This is achieved by arranging files so as to fill removable disk packs as much as possible, at the same time remembering the need for expansion and overflow. The employment of fixed disks is best limited to applications in which the files may be retained permanently on the disks. The amount of data held on fixed disks is generally so large that it is not a viable proposition to load and unload the disks except from high-speed magnetic tape. The combination of fixed disks and high-speed tapes results in an expensive configuration.

*Peripherals and terminals*

The type(s) of input peripheral depends largely on the data preparation/entry problems. Having decided upon a key-to-disk system, the computer input is magnetic tape and this presents less of a problem than other media. With punched cards, paper tape, document readers etc., the speed of the peripheral is significant. The decision may fall between one fast unit or two or more slower ones. More than one unit provides a higher degree of security against failure but this is relatively costly and ineffective for sequenced input.

Output peripherals are dominantly printers (Section 10.6). It is worth giving considerable thought to the type and speed of printer(s) to be used. Line printers are faster, costlier and bulkier than serial printers. Generally a chain printer gives a higher quality of print than a drum printer, and the removable chain enables various print founts to be used. The more recent types of serial printers, such as ink-jet and electrothermal, demand careful consideration before committing them to extensive output requirements. Their technology may not be sufficiently developed to cope with long periods of continuous printing.

A wide range of VDUs is available and so the systems analyst can make a scrupulous choice in regard to display capacity, character size and other features (Section 10.6). The VDU differs from other units in that it interfaces more closely and continuously with people, and so a model's suitability in this respect needs consideration. If an inter-active system utilising man-computer dialogues [10.35] is planned, the keyboard, controls and features need to be carefully examined. Particular aspects are removable keyboards to suit different categories of users, split screening for different jobs, and the special display features for the various types of output. Visual display is developing quite rapidly and so the systems analyst must watch for new tech-nology such as plasma display and coloured display in case these are beneficial to his system.

## 11.6   References and further reading

*Decision tables*

11.1   *Decision tables in data processing,* NCC (1970).
11.2   'A basis for decision', *Data Systems* (May 1970).
11.3   'Decision table translation', *Computer Bulletin* (December 1969).
11.4   'The use of decision tables within SYSTEMATICS', *Computer Journal* (August 1968).
11.5   'The interpretation of limited entry decision table format and relationships among conditions', *ibid.* (November 1969).
11.6   'A new rule mask technique for interpreting decision tables', *Computer Bulletin* (May 1969).
11.7   'A rule mask technique for decision table translation', *ibid.* (October 1970).
11.8   'Decision tables as a systems technique', *Computers & Auto-mation* (April 1970).

11.9    *Recommendations for decision tables used in d.p.*, British
        Standard BS 5487 (1977).
11.10   'Converting decision tables to computer programs' *Computer
        Journal* (May 1975)


*Timing*

11.11   'Computer timing and costing model', *Data Processing* (July-
        August 1968).
11.12   'System for timing systems', *Computer Bulletin* (November
        1968).


*Auditing*

11.13   CLIFTON and LUCEY, *Accounting and computer systems*,
        Chapter 9, Business Books (1973).
11.14   *Internal control and audit of computer-based accounting
        systems*, Institute of Chartered Accountants (1970).
11.15   'The auditor and electronic data processing', *Computer Jour-
        nal* (July 1965).
11.16   'Computer audit packages', *Computer Bulletin* (June 1970).
11.17   'Auditing magnetic tape systems', *Computer Journal* (April
        1964).
11.18   WASHBROOK, *Management control, auditing and the com-
        puter*, Heinemann (1971).


*Computer configuration*

11.19   'A review and comparison of certain methods of computer
        performance evaluation', *Computer Bulletin* (May 1968).
11.20   CHORAFAS, *Selecting a computer system*, Gee (1970).
11.21   SMYTHE, *Choosing a computer*, Business Publications (1971).
11.22   'Choosing computers for the post office', *Computer Bulletin*
        (March 1967).
11.23   'When is a terminal a computer?', *Data Systems* (June 1969).
11.24   'Maxiterminals and minicomputers', *ibid.* (January 1970).
11.25   'Terminal based systems', *ibid.* (January-February 1974).

# Chapter Twelve
# *Appraisal of data processing systems*

## 12.1  Comparison with existing system

When the new data processing system has been designed, the systems analyst should evaluate it in comparison with the system that it is intended to replace. This comparison should cover not only the advantages of the data processing system over the existing system, but also an honest assessment of any disadvantages. By making himself aware of these advantages and disadvantages, the systems analyst is able, when presenting the new system to the management and staff, to forestall criticism of it.

The comparable aspects of the new system and the existing system are financial, speed of throughput and quality of information produced (Section 12.2).

*Financial comparison*

The existing system will have been cost estimated previously (Section 4.8), and the costs of the proposed new system are now estimated bearing in mind that they will not be constant from year to year.

The financial factors that enter into the installation of a data processing system are equipment, environment, materials, staff and training costs.

*Equipment costs*   Quite apart from the range of equipment involved in the system, a prime consideration is the method of purchase. The merits of outright purchase, rental, leasing, or other financial arrangements ought to be studied by the financial executives of the organization, guided by the analyst's advice on the hardware aspects. The costs of capital equipment include not only those for the computer and its peripherals and terminals, but also some or all of the items listed below:

Communication equipment.
Card punches and verifiers.
Data recorders, e.g. key-to-disk systems.
Stationery handling equipment (decollators, bursters, etc.).
Extra reels of magnetic tape.
Extra removable disks, magnetic ledger cards, etc.
Trays, trolleys, racks and cabinets for cards, magnetic tape and disks.
Additional furniture and fittings.

*Environment costs*  These include all factors connected with the preparation of the computer room, and sometimes this involves the erection of a new building or large-scale constructional changes to an existing site. In any case even if an existing room is to be used, some or all of the following will be needed:

Air conditioning plant and associated control gear (capital and maintenance costs).
False flooring or other arrangements for covering cabling.
Acoustic and thermal insulation including the double glazing of outside windows.
Wall and ceiling preparation.
Fire prevention equipment.
Other extra accommodation such as d.p. staff offices, store rooms, ancillary machine room, data entry/preparation room, maintenance engineer's room, air conditioning plant room.
Any special arrangements for moving equipment in and out of the building.

With real-time systems using widespread terminals, the above costs are amplified very considerably. Not only is the hardware much more extensive but it is also necessary to allow for the costs of data transmission — either on a line rental basis or possibly the installation of private lines. The environment costs are also likely to be higher because of the increased amount of hardware at the central computer site.

*Running costs*  These are fairly high at first but continue at a lower level throughout the life of the system; they include:

Equipment maintenance.
Punched cards, stationery, print ribbons and other consumable materials.
Insurance of equipment.
Stand-by charges.

Replacement or reconditioning of disks and tapes.
Books, manuals and other literature and stationery.
Air conditioning filters and materials.

*Staff costs*   These include all members of the data processing department, and also the staff retained in other departments who are connected with the data processing system and who were included in the existing system costs. The data processing department staff consists of:
Manager(s).
Systems analysts.
Programmers.
Computer operators.
Data entry clerks.
Control staff (tape librarian, control clerks).
A handyman (principally for moving materials).
Computer maintenance engineer.

When a large number of on-line terminal operators are involved, it is best to estimate their costs on a separate basis since these may be very considerable. It is generally the situation however that these people are existing employees put to a new role so that there is not necessarily any significant net increase in their cost.

*Development costs*   These are high initially and continue to a small degree dependent largely upon the rate of staff turnover. The initial training costs are made up of:
Software purchases.
Consultants' fees.
Changeover and external program testing.
Training, accommodation and travelling expenses of trainees.
Terminal operator training and subsequent computer training mode facilities.

*Comparison of speeds of throughput*

'Time is money' – but this is not relevant if it is someone else's time. When comparing the throughput speeds of the new system and the existing system, the significant point is what benefits are accrued from the reduced throughput time? These may be directly measurable, for instance as a result of being able to collect debts earlier; or they may be intangible, such as improved customer service. Intangible benefits are difficult to detect let alone measure, but if the improved service is of sufficient magnitude, its effect will show up in later sales figures. Whenever possible an estimate should be made of intangible savings.

Other benefits of reduced throughput time might well be:

1   Reduced waiting time in the factory, this benefit being obtained by faster re-planning.
2   Quicker detection and reporting of situations that require human action.
3   More up-to-date management reports.
4   Better creditor payment control – the faster throughput of creditor payments after authorisation means that they can be delayed until the most propitious time from the financial aspect.

*Cost comparison table*

This is a convenient method for demonstrating the comparative costs and savings; it is suitably drawn up after the style of Figure 12.1 The reduction in existing work may be estimated from the Staff Utilization Table (Figure 4.6), and the financial savings estimated from the Activity Costs Table (Figure 4.7).

Any government grants or allowances are, of course, offset against the data processing costs. Both the additional costs and the potential savings are calculated annually to cover a sufficiently long period of time for the system to settle down. The example in Figure 12.1 covers a five-year period and the question arises as to the present-day value of these future costs and savings. A single-figure measure of a set of future amounts is obtained by calculating their 'internal rate of return' (IRR). This is a percentage figure that is equivalent to the interest returned on invested capital. The annual net costs are regarded as investments and the annual net savings as returns. The IRR of a computer system must be at least as great as the return that could be achieved by investing the capital in other ways.

In order to calculate the IRR, we have to find a percentage which, when used to discount future amounts, brings the sum of the discounted amounts to zero. Putting this mathematically for the example in Figure 12.1b, where $x$ is the required IRR:

$$-60,790 - \frac{8950}{(1+x)} + \frac{60,750}{(1+x)^2} + \frac{80,750}{(1+x)^3} + \frac{88,750}{(1+x)^4} = 0$$

Solving this equation gives $x = 0.514$, thus the example is equivalent to investing capital at an interest rate of 51.4 per cent.

It is not easy to solve polynomial equations such as this; the simplest ways are either to use an investment appraisal package on a computer or to 'guestimate' the value of $x$. The guestimate of $x$ is increased if the equation's total turns out to be positive, and vice versa, until a value of $x$ is found that makes the total virtually zero.

## 12.2 Management information and control

The paramount consideration in regard to the management information provided by the data processing system is quality. This means its accuracy to the degree necessary, its topicality, its conciseness and its comprehensiveness. The analyst endeavours to design a system that provides all these characteristics in its output of management information, and in so doing finds it necessary to strike a balance between conciseness and comprehensiveness. How can these two apparently conflicting aims be reconciled? He must not present management with a mass of detail such as is only too easy to produce with a computer. On the other hand whatever information is prepared must be based on all the facts of the situation.

*Exception information*   A solution to this problem lies in the provision only of information that leads to action or decision on the part of management; this information is, however, based upon the full complement of relevant data. This principle is termed 'management by exception', and the skill in design lies in deciding what is exceptional. Management will have provided some indication of their information requirements when they were interviewed during the investigation, but these requirements may disappear as a result of automatic action by the data processing system. It is therefore prudent to make further inquiries of management during the design stage, and at this time they can be told the general line of approach of the system so as to guide them in redefining or re-affirming their information requirements. Every piece of information has its value, although this value is not easily measured in financial terms. Nevertheless the systems analyst must make some evaluation of required information so that this can be compared with the cost of producing it. It is obviously unsatisfactory to allow a manager's whim to instigate an expensive piece of hardware or a time-consuming computer routine.

The same principle applies to the degree of control that is aimed at; the cost of any improvements in this respect must be commensurate with their value to the company. It is, for instance, pointless to introduce a special checking routine that costs the company £1000 per annum, when the savings made through the checking are known never to exceed £900 per annum.

*Management information reports*   These are required for four main reasons:
1   *Action*   Often associated with unforeseen circumstances that call for emergency action, this normally involves middle-level management rather than top management.
2   *Assurance*   That operations within a department or factory are

DATA PROCESSING — ADDITIONAL COSTS £

| Item | Type of cost | Financial year | | | | |
|---|---|---|---|---|---|---|
| | | 1980-1 | 1981-2 | 1982-3 | 1983-4 | 1984-5 |
| Computer | R | 34500 | 34500 | 34500 | 34500 | 34500 |
| Key-to-disk | P | 5500 | 1100 | | | |
| Maintenance | R | 250 | 300 | 300 | 300 | 300 |
| Tape reels | P | 900 | 250 | | 100 | 100 |
| Other equipment | P | 200 | 50 | 50 | | |
| Equipment totals | | 42490 | 36700 | 35100 | 35150 | 35150 |
| Air conditioning | P | 7500 | | | | |
| Air conditioning maintenance | R | 150 | 150 | 150 | 150 | 150 |
| Environment totals | | 9200 | 200 | 200 | 200 | 200 |
| Maintenance | R | 1400 | 800 | 800 | 800 | 800 |
| Stationery | R | 450 | 100 | 100 | 100 | 100 |
| Running totals | | 2000 | 950 | 950 | 900 | 900 |
| D.P. mangt. salaries | R | 3500 | 3500 | 3500 | 3500 | 3500 |
| Systems salaries | R | 4900 | 4900 | 4900 | 4900 | 4900 |
| Progrs. salaries | R | 7000 | 9000 | 9000 | 9000 | 9000 |
| Salaries and o/h totals | | 18500 | 21000 | 22000 | 22000 | 22000 |
| Training | R | 1200 | 400 | 300 | 300 | 300 |
| Misc. costs totals | | 2000 | 1100 | 1000 | 1000 | 1000 |
| Grand totals (costs) | | 74190 | 59950 | 59250 | 59250 | 59250 |

P = Outright purchase (capital) cost,
R = Rental or regular cost.

**Figure 12.1a**  *Costs of new system*

DATA PROCESSING — POTENTIAL SAVINGS £

| Dept. | Routine | 1980-1 | 1981-2 | 1982-3 | 1983-4 | 1984-5 |
|---|---|---|---|---|---|---|
| Stores ctl. | Filing and sorting | 7000 | 14870 | 14870 | 14870 | 14870 |
| | Ledgers | | 5000 | 13299 | 13299 | 13299 |
| | Stock checkg. | | | 5500 | 5500 | 5500 |
| | Purchase reqs. | | 2500 | 6500 | 7500 | 7500 |
| | Works orders | | | 5000 | 5000 | 5000 |
| | Other work | 400 | 1100 | 1800 | 2000 | 2000 |
| Stores ctl. office totals | | 7400 | 23470 | 46969 | 48169 | 48169 |
| Purchasing | Ledgers | | 3600 | 5800 | 5800 | 5800 |
| | Typing | | 2450 | 4800 | 4800 | 4800 |
| Purchasing totals | | | 9000 | 13000 | 22500 | 22500 |
| Production ctl. totals | | | | 5500 | 12000 | 18000 |
| Wages dept. totals | | 6000 | 10000 | 10000 | 10000 | 10000 |
| Sales invoicing totals | | | 4000 | 8000 | 11000 | 11000 |
| Grand totals (savings | | 13400 | 51000 | 120000 | 140000 | 148000 |
| Less grand totals (costs) | | 74190 | 59950 | 59250 | 59250 | 59250 |
| Annual net savings | | −60790 | −8950 | 60750 | 80750 | 88750 |

**Figure 12.1b**   *Net savings of new system*

197

proceeding, within accepted limits, according to plan.

3    *Tactical planning*   The comparatively short-term disposition of the manager's resources, such as machine shop loading for the following week.

4    *Strategic planning*   The longer-term planning of future markets and products, and deployment of financial resources.

## *Management information considerations*

In designing a management information system the systems analyst will always be torn between providing everything that the manager requires by way of information and keeping the system straight-forward and economic. As suggested earlier, management information appears in general to be disproportionately expensive as compared with other, more mundane, computer output. This is even more so when the required information is changeable from day to day, and the systems designer is inevitably forced to prune the information asked for. Unfortunately the pruning is not easy if we are to be sure that nothing vital is lost and also that nothing trivial remains at high cost.

There are relatively few business situations that justify the employment of interactive terminals purely for management's benefit. Interactive computing can provide immediate information of a wide variety and flexibility but only at high cost in hardware and tremendous effort in planning and programming. Essentially the systems analyst must obtain answers to the following questions before committing himself to designing a management information system involving costs substantially above those already incurred for basic data processing.

1    Does the manager asking for information appreciate its cost?

2    To what extent is this requirement of a temporary nature?

3    Are the parameters of the information likely to change and can all of these be pre-determined?

4    How does each manager's information needs fit in with those of the other managers? This is particularly pertinent to the unintentional replication of information in different reports.

5    What is the time factor involved, and if immediate is there sufficient justification to warrant a real-time or on-line system?

6    If the management information requirements are obscure, what additional data should be held on files to cater for these when they become clear?

Although it is improbable that all the above questions are answerable unambiguously, their posing in itself stimulates management into analysing their demands more critically — this can do nothing but good!

## 12.3 Data processing facilities

In Chapter 11 the problem of determining the computer's configuration was discussed. It must not be assumed however that the purchase of the user's own computer is necessarily the best means of implementing data processing in all situations. There are a number of alternatives to this, and they should all be given due consideration. The alternatives to outright purchase of the computer are renting, leasing, hire purchase or other financial arrangement agreed with the supplier. Each of these has its advantages with regard to taxation, depreciation and government allowances. From the systems design and operating aspects, the above arrangements are identical since they all involve the computer being on the user's own premises. The main alternative facilities to this are:

    Service bureaux (data processing or computer bureaux).

    Time hire.

    On-line time-sharing.

    Shared computer.

*Bureau service*

This facility is commonly employed by a user as a means of experimenting with data processing. Its principal advantage is its comparatively low cost, usually as a result of the limited scope of the work carried out. A common arrangement is for companies to have a few restricted applications done by a service bureau prior to installing their own machine. In other cases an application is chosen for permanent implementation on a service bureau basis because of its special difficulty by other means, or its large volume of data. The application may be one in which the bureau specializes, such as PERT, stockbroking or spares control. Most bureaux accept work of a general nature and create tailor-made programs to meet their client's precise requirements.

*Time hire*

The user hires a computer for a certain length of time at pre-booked regular or irregular intervals. At best this is equivalent to the user having a computer of his own for a restricted time. At its worst it is a tedious and time-wasting method of obtaining the desired results. The degree of efficiency obtainable with this facility depends upon the hired computer's configuration, its distance from the user's premises, and the planning of its hire and use in relation to data preparation and other user activities.

## On-line time-sharing

With this facility the user shares the computer simultaneously with a number of other users. This is accomplished by the user's premises containing a direct link to the computer in the form of a terminal, and data is transmitted to and from the computer over communication lines. A time-sharing computer must be sufficiently powerful to handle the data from all the terminals rapidly.

Thus, because the computer's response is immediate, the user is not aware that he is sharing the computer simultaneously with others. Each terminal is equipped to meet the particular needs of its user, including fast peripherals and intelligence.

Another arrangement is for the user to transmit input data via his terminal but for the output to be returned by post, e.g. sales invoices.

## Shared computer

This is not to be confused with time-sharing; a shared computer merely means that several users combine to purchase a computer and then apportion its working time between them. The users are often members of the same group or association. The computer may either be in a data processing department staffed by 'group' employees, or each user has his own data processing staff for programming and operating. These two arrangements are in effect the same as bureau service and time-hire respectively.

## Factors in choice of facility

There are obviously many factors that can enter into a particular company's decision regarding the choice of facility and the applications thereon. Some of the factors relate purely to the company's circumstances at one point of time or to its long-term plans. The three main general factors are, however, cost, staff involvement and file accessibility.

*Cost*  This is generally of prime importance when making the choice, but the systems analyst must be certain that he is comparing like with like. It is, for instance, unlikely that the employment of a service bureau can be properly compared with having one's own computer as the former cannot attempt an integrated or total system as far as most users are concerned. A bureau functions best when employed for a few clear-cut applications. The other facilities are more comparable with each other since they can all achieve the same results.

When comparing the quoted or calculated costs of the work, it

200

must be remembered that allowance is necessary for contingencies such as re-runs. And that in the case of time hire, and possibly a shared computer, additional costs are introduced by the transportation of input media, output results and the operating staff themselves.

*Staff involvement*   The odd man out is the batch processing bureau service; with all the other facilities the user's own staff are deeply involved in planning and operating the system but in the case of bureau service this may not necessarily be so. If it is the intention to employ a bureau to implement an isolated and straightforward application, which is an end in itself, there is no need for deep involvement of the user's staff. Nevertheless at least one member of staff should be made responsible for maintaining close contact with the bureau, and keeping himself informed about the work in hand. This may sound obvious, but it is a fact of life that some bureau users have lost real contact with the bureau, with the consequence that the results they receive have become almost meaningless to them.

If the bureau service work is at all comprehensive or complete, it is essential that the user's systems analysts are effectively in control of the system. They should be of the same calibre as for internal computer usage since they are equally responsible for designing the system. It is both unwise and unreasonable to expect the bureau's staff to do detailed or comprehensive systems investigation and design for the user. Neither the charges involved nor the bureau's staff establishment are likely to be such as to make this a viable proposition except in isolated cases.

If a user wishes to employ a service bureau without having his own systems analyst, the answer is to commission a computer consultant to act in this capacity. Care must be taken when choosing the consultancy firm that its staff have practical experience of actually implementing work on a computer.

In some cases a consultancy firm or service bureau provides a 'turnkey operation' for their client. This implies that the consultants carry out the complete investigation, design and implementation of a new data processing system, with a minimum of involvement on the client's part. It is then handed over to the client for operating.

Another arrangement available to a prospective computer user is 'facilities management'. This means the establishment, management and operation of all the client's data processing activities by a bureau. Facilities management, if truly implemented, is tantamount to the bureau staff being employees of the client since they must obviously become deeply involved in and committed to the client's interests.

In practice it is unusual for a turnkey operation or facilities management to be fully implemented, and in any event the legal and financial connections between the client and the bureau need to be well established.

*Accessibility to files*   In this context accessibility means the lack of remoteness of the files. Files that are stored in a user's on-site computer are obviously the least remote and the most readily accessible. An on-line time-sharing system also provides good accessibility, but the other facilities may not do so. This lack of file accessibility is often of no consequence for isolated applications, but it introduces great difficulties for the more comprehensive systems. The problem is related to both the frequency of the need to have access to the files and to the intervals between the user's contacts with the computer. In circumstances where file information is required at short notice, printed copies of the files have to be prepared at regular intervals and held on the user's premises. Although this procedure may be satisfactory for low activity and non-volatile master files, the danger arises with other files that the printed copies are out-of-date when consulted.

## 12.4   System presentation and management decisions

After the design and appraisal of the data processing system, it should be formally presented to top management before implementation commences. Depending upon the decisions made at the feasibility stage, the presentation might involve top management in arriving at the decision whether or not to proceed with data processing. In any case, even when the decision to proceed was made earlier, the sum of money and the re-organization inherent in this step make it vitally important that top management are fully aware of its consequences and benefits. It is also probable that although the decision to go ahead has been made previously, the precise equipment and/or facility have not been chosen.

It is unfortunate but nevertheless true that a computer is one of the most difficult of machines to really understand and yet one of the easiest to be impressed by. Although, in the final analysis, top management sanctions the order for a computer, the decision as to its identity is for the specialist to make. Only the company's data processing staff are qualified to arrive at the computer's configuration, and their opinion in the selection of the particular model should carry the utmost weight. In special circumstances, for instance due to inter-trading, government pressure or tax allowances, top management may decide to place the order with a certain manufacturer. Nevertheless the configuration of the computer should be specified by the user's systems analyst and not by the manufacturer.

What can the systems analyst do in order to give management a clear picture of the situation? As mentioned in Section 9.5, a systems definition ought to be prepared in the design stage. This definition is intended not only for the staff who will implement the system, but some sections of it are suitable for assimilation by non-technical staff such as top management.

*Discussions*   These are useful to the systems analyst for putting forward ideas about the proposed system, and for receiving criticism of it. Having listened to critical comments during the discussions prior to the formal presentation of the system, they can be disposed of either by adjustments to the system, or by further explanation. Discussions help to engender a feeling of participation among top management, and they also encourage the self-styled ignoramus to try to understand the system.

*Talks*   These are given by the systems analyst for the benefit of all levels of management, and should be brief, informal, informative and interesting. Their main aim is to make management feel that the new system is a step towards the ideal means of control, and that it can be developed logically in this direction.

Certain things are better explained verbally than in writing, and in this category are the deliberate omissions from the system and from the recommended hardware. Management does not like to feel that their associates or competitors have more advanced ideas, and these talks provide an excellent opportunity to explain not only the proposed system and hardware, but also the advantages of these in comparison with alternatives. For instance, visual display units may have been installed by an associated company and have taken the managing director's fancy; a rational explanation that these have been considered but found unsuitable, puts his mind at rest.

The talks can also be used to overcome any points of apprehension that management may have about the proposed data processing system. These are generally connected with security, so that assurances about this aspect are well worth while.

*Management decisions*

Having been formally presented with details of the proposed system, and if not already decided, top management must, after due deliberation, arrive at a decisive conclusion. This may be *(a)* to go ahead as per the recommendations, *(b)* to make amendments to the system's output or objectives, *(c)* to postpone the implementation of the system for a stated period, or *(d)* to reject the proposed system. Conclusion *(b)* results in the reconsideration at a later date of the amended recommendations. The reasons for conclusions *(c)* and *(d)* should be made known although some reasons, such as a pending mergers, take-overs or top-level policy changes, may make this difficult.

The situation to be avoided is disinterest or permanent indecision on the part of top management; this is not only demoralizing for the systems analyst, but also leaves a feeling of apprehension lingering in the minds of other staff.

## 12.5  References and further reading

*Finance and costs*

12.1  MERRETT and SYKES, *Capital budgeting and company finance,* Longmans (1969).
12.2  'Financing the computer', *Data Processing* (November-December 1969).
12.3  'Paying for your computer', *ibid.* (October 1976).
12.4  'Computer consumables', *ibid.* (September-October 1968).

*Management information*

12.5  'Management information systems − can computers help?', *Computer Bulletin* (March 1968).
12.6  'A survey of management information systems literature', *ibid.* (June 1971).
12.7  *The computer as an aid to management,* Institute of Chartered Accountants (1968).
12.8  'Information retrieval for management', *Data Systems* (July 1967).
12.9  *Management information systems − annotated,* Institute of Chartered Accountants (1969).
12.10  'Management information retrieval', *Computer Journal* (May 1970).
12.11  'Answering the questions of top management', *Computer Weekly* No. 186 (9 April 1970).
12.12  'Modular method of structuring MIS', *Journal of Data Management* (February 1970).

*Bureaux facilities*

12.13  *The use of computer service bureaux,* Cert. Acc. Educ. Trust (1973).
12.14  'That bureau business', *Data Systems* (February 1971).
12.15  'Service bureaux or a visible record computer?', *Computer Management* (November 1970).
12.16  'The rise of the bureaux', *ibid.* (September 1970).

12.17 'Computer bureaux – the way ahead', *Computer Bulletin* (March and April 1972).
12.18 'Services of the future', *Data Systems* (June 1970).
12.19 'Computer bureau and consultancy services', *Data Processing* (May-June 1969).
12.20 'Do OCR bureaux have a future?', *ibid.* (March-April 1971).


*Time-sharing*

12.21 'Criteria for remote intelligence', *Computer Management* (July 1972).
12.22 'A user's view of the U.K. time-sharing industry', *Computer Bulletin* (January 1970).
12.23 'The impact of multi-access', *ibid.* (March 1968).
12.24 ZEIGLER, *Time-sharing data processing*, Prentice-Hall (1967).
12.25 'Time-sharing explosion', *Data Systems* (June 1969).
12.26 'Time-sharing, an accepted technique', *Data Processing* (January-February 1969).

# Chapter Thirteen
# *Implementing data processing systems*

## 13.1    Systems testing

Before bringing the data processing system into use, it is of vital importance that it is both comprehensive within its intended limits and fully correct. Each program will by now have been written according to its specification (Section 10.8), and tested by the programmer to his complete satisfaction. The final responsibility for the correctness of both the suite of programs and the system as a whole lies with the systems analyst. He must make absolutely certain that each run produces exactly what is required of it, and that the runs link smoothly together to provide the correct output of each routine. The linking together of runs during the testing procedures ensures that the output of one run is entirely correct as the input to the following run. The linking of runs is often bedeviled by the fact that different programmers have written the programs in a routine, and communication between them about technical details has not been all it could have been.

Ideally one set of basic data can be used to test all the programs in a routine. In practice this method is usually unsuitable because test data chosen to extend one program's capabilities does not necessarily do so for another; thus there is often a conflict between the requirements of the respective data. The programmer will have created his own test data; the systems analyst should create another quite separate lot so that a double check is made.

## Characteristics of test data

1   It must extend the run into its limits with regard to factors such as:
a   The sizes of input and output fields.
b   The sizes of calculated intermediate values.
c   The variations in the possible formats of indicative fields, e.g. code numbers.
d   The detection of errors by feasibility checks, and the consequent action by the computer.
e   The variations of program paths – although it may not be possible to test all combinations of these at one time.
f   The interpretation of special symbols and coded data items.

2   The capacity of storage areas allocated to tables, indexes, files and results should be tested as far as is reasonably possible. This is particularly applicable to direct access files in which overflow of records is expected.

3   The testing of data organization is generally more difficult than the checking of calculations; two separate lots of test data are therefore probably needed. Data organization testing is especially relevant to the addressing of records and the overflow techniques used in direct access files.

4   Not only the visible output but also the data left stored on magnetic tape or disks at the end of a run needs checking. Although this usually forms the input of another run, this is not really a suitable way of checking it. A special print-out of this stored data, followed by a manual check, is necessary.

5   It does not follow that 'live' data is necessarily the most suitable for test purposes; the reasons for this are as follows:
a   Live data is almost sure to be biased towards certain characteristics, for instance a predominance of certain code numbers or the absence of certain types of data.
b   The output results from the previous usage of the live data were often not intended to be exactly the same as from the new system. The reconciliation of the different results introduces more work than the use of fresh data.
c   Live data does not necessarily extend the program in the ways mentioned in 1, 2 and 3 above.

6   Test data is also used to check that run times approximate to those that have been estimated. Live data is convenient for this purpose because in this respect it is representative of that which will actually be processed by the run.

One of the problems intrinsic to systems testing is that of providing enough data to form pseudo master files for program testing purposes. If it is possible to create the actual master files before testing the other parts of the system, then the problem disappears. This is not always convenient however, and so dummy master files have to be created. Instead of devising and entering a large amount of artificial data, it is quicker and cheaper to create it by means of a 'data generating' program. This is written in a generalized form and set up by parameters to suit the particular type of record that is required. The factors involved in data generating are:

1    Each data item in a record is of a given format and lies within maximum and minimum limits.
2    Each record consists of a given set of data items, some of which are fixed in number and others are variable within specified limits.
3    The records may be stored either sequentially according to a given key, or randomly, and are assigned to their respective locations in a direct access file by the normal file creation program (Section 13.2).

The data generating program creates the data items of a record one at a time by using generating pseudo random numbers. After being generated the pseudo random number is checked for format and size; if acceptable it is then inserted into the record to represent a data item. This is repeated for each data item in the record according to its parameters, and when complete the record is written away to the file. If the records are being created for a sequential file, the keys are generated and inserted, and the records then sorted into sequence before being written to the file.

## 13.2   Creation of master files

As suggested earlier, the master files are the framework or database of a data processing system. It is therefore essential that they are created initially in a complete and accurate form. If this is not done it is possible that some of the errors could remain undetected for a considerable time, with consequent errors in the output of the routines. In the more straightforward situations the master file is created by reading cards and simply copying the data on to magnetic tape or exchangeable disks. In other cases the routines for file creation are complicated, even to the extent of being some of the most difficult jobs to program and execute.

*Sources of master file data*

It is quite usual for a number of different source documents to contribute data to the master file. These documents may well have been
used in different departments of the company for many years, but
may never previously have been brought together. This means that a
lack of uniformity between documents almost certainly exists,
including differences in data items such as descriptions and code
numbers. An example of this state of affairs could be in the creation
of a master file for sales/stock processing; the stock levels being
derived from the stock sheets, the sales history from the sales record
cards, and the selling price from the official price list. Because of the
diverse nature of the previous applications of these documents, their
non-uniform code numbers have been of no consequence. The inter-
connection of these documents was formerly through manual procedures so that discrepancies were dealt with by human memory and
intuition.

It is commonly the case that items are present in one set of documents and missing from another set. This situation raises difficulties
in the later stages of file creation if the file data is to be drawn from
both sets; it is therefore essential that matching and correction runs
are incorporated in the early stages of file creation routines.

*File creation from sub-files*　When a number of sources provide data
for a master file, it is occasionally possible to combine the documents
manually prior to keying or punching the data as a set for each record.
This is however a tedious process both as regards manually collating
the documents and keying from them. It is often safer to keep the
documents separate and to create a sub-file of cards or tape from
each lot of similar documents. Each sub-file is transferred on to magnetic storage and, at the same time, a proof list is prepared and control totals accumulated and printed. The sub-files are then sorted and
merged to form the master file; any records for which data is missing
are omitted from the master file. Using the list of incomplete records,
the missing data is found and inserted during a subsequent run.

An alternative method is to create the sub-files as above, but then
merely to match the records without actually merging into one file.
The result of this run is a printed list of omissions which is used to
find the missing documents. The sub-files are then corrected
separately by insertion of the omissions, and finally merged to form
one master file.

*Proof lists*　The manual checking of a full detailed list of items in a
large file can prove to be an overwhelming task. Nevertheless it is
useful to have a proof list available for checking individual items as
and when necessary. With descriptive proof lists, it is sometimes

possible to accelerate the checking process by merely scanning the items and checking against the source documents only those that appear untoward.

*Control totals*    These are easily obtained from a computer run but not always so from a manual procedure. In spite of this difficulty it is worthwhile to obtain control totals from the source documents along the lines described in Section 11.4. If, as is likely, they do not appear as part of the existing system, they should be created by using an adding machine for batches of not more than 2000 documents at a time. Thereafter the control totals are accumulated and checked at every stage of the file creation procedure.

*Time spread of master file creation*

It is frequently the case that a master file has to be created from source documents that are being continuously amended. This is not so much the situation with descriptive data as with stock, sales and production figures, for which there is little chance of finding a static situation. If master files could be created instantaneously, there would be little problem; but since they usually take a considerable time to create, the data for one record tends to appertain to a different data to that for another. The essence of this problem is connected with the initial entry of data, remembering that once this has been done for a source record, all further amendments to that record must also be entered later so that it is absolutely up-to-date when the master file comes into use. Great care is necessary to ensure that no amendments nor transactions are overlooked or duplicated after the date on which the initial record is created.

A somewhat expensive method of achieving this end is to make photocopies of all documents as at a certain date and time; data entry from these can then proceed at leisure and later amendments entered at the same time. A more economical method is to enter a batch at a time, in one run outside normal working hours. This means that no changes can occur during the time of entry, and amendments to each batch are clearly applicable from the date of entry. As an extra safeguard, each record can initially contain its date of creation so that, provided subsequent amendments and transactions are also dated, there will be no doubt as to their authenticity for master file updating.

## 13.3   Changeover procedures

The changeover from the existing system to the data processing system begins after the computer has been installed, but preparation for the changeover will have been going on for some time before this event. System testing should commence on other computers well before the installation of the user's own machine so that useful results can be produced immediately after its installation.

There are four basic procedures for achieving the changeover; the one that is adopted in a particular situation depends upon the type of organization and the relationship of the old and new systems. Different procedures may be employed for the various applications within the one company, and it does not follow that the same procedure should necessarily be employed for the same application in different companies. The most suitable procedure is chosen by the systems analyst according to the prevailing circumstances.

*Direct changeover*

This involves ceasing operations under the old system and commencing the new system immediately afterwards. This is a somewhat drastic method and should be adopted only if no other procedure is suitable, for example, when it is impractical to meet the demands of extra work entailed in the other procedures. Whenever possible, direct changeover should be fitted into a weekend or statutory holiday period when work is slackest. This causes the minimum of disruption and allows the most time to bring the new system into operation.

One of the main disadvantages of direct changeover is the problem of ensuring that the new system is functioning one hundred per cent correctly. There is no basis of comparison with the previous system because the same results have not been produced previously. Also if results are found to be incorrect, it may be difficult to retrieve the situation, especially if the old system cannot be re-employed temporarily until things are put right with the new system.

In the case of most terminal-based real-time systems, such as for example banking, direct changeover might well be the only possibility. A modification of this is step-by-step changeover in which a routine of application is transferred at a time on to the new system. This latter arrangement is satisfactory provided there is a suitable and clear division between the routines and no difficulties occur in regard to file creation.

Direct changeover needs the most careful planning by the systems analyst, and the most diligent attention to timing by the operational staff. Given these requirements and a little good luck, the task can be accomplished successfully.

*Parallel running*

With this procedure the current basic data is processed by both the old and new systems and, as far as it is practicable, the two lots of results are checked against each other. This is generally done by comparing all totals individually, and making sample comparisons of the detail if the amount of output is extensive. Care must be taken to ensure that not only the printed output of the new system is correct but also the carried-forward data.

Parallel running is normally carried out for one or two processing cycles, after the first of which the output from the old system is distributed; thereafter the new output is distributed and the old output held in reserve in case of dispute. In cases where the new output has a very different format to the old, both sets of internal documents may be distributed together on the first occasion, if necessary accompanied by an explanatory note pointing out the changes. This arrangement is not advisable with externally distributed documents, but a brief explanation may be worth sending with the new documents in some circumstances.

The major problem with parallel running is the duplication of work involved. In all probability the staff who have been working on the old system will also be involved in the new system; for this reason parallel running must not be more prolonged than is absolutely necessary.

*Pilot runs*

In this procedure current data continues to be processed by the old system while previous data is re-processed by the new system. Thus, provided the two systems are intended to produce similar results, a basis of comparison exists. The amount of previous data re-processed by the new system depends upon the particular application; it is not usually necessary to process all of a cycle's previous data but care must be taken to ensure that the chosen sample is truly representative. Provided the sample is not too extensive, there is no great hardship in pilot running for several processing cycles. If an application consists of sections of data that have different characteristics, each such section may be used as pilot data in a different cycle. By this means each section gets a check that is directed towards the particular characteristics of its data.

When the new system is proved to be correct, a double cycle in one period makes the pilot run into a parallel run. Thereafter the old system can be abandoned in the knowledge that extensive checking of the new system has been carried out.

*Phased changeover*

This is similar to parallel running except that initially only a portion
of the current data is run in parallel on the new system, for instance
that appertaining to one department or section. During the following
weeks more sections are transferred on to the new system, and in
each case the old system runs in parallel for one processing cycle only.
Thus the old system is phased out as the new system builds up, and
at each stage it is quite practical to check the new output against the
old before distributing it. The total amount of extra work is generally
less than that involved in parallel running.

## 13.4   Involvement of user department staff

The data processing department cannot function in a vacuum; because
it is essentially a service department, it must be employed by and be
in contact with the other departments in the organization. The
greatest benefit is derived from the data processing system when
other staff are not only aware of its existence but have a genuine
desire to make use of its services. This atmosphere is best created by
the systems analyst actively 'advertising' the data processing system
during its appraisal and implementation stages. This is less necessary if
project teams containing user department staff were used in the
systems investigation and contact was maintained thereafter.

The staff from outside the data processing department can be regarded as falling into one of three groups:
1   Those who will be feeding data into the system and/or receiving
    results from it.
2   Management receiving information from it.
3   Other staff, whose work will be indirectly affected by the
    system.

These people, together with the organization's outside contacts,
form the environment of the data processing system. Once the
system is in operation, it may be too late to modify the environment
if it proves to be confused or uncooperative. The systems analyst is
well advised to encourage the right attitude among other staff prior
to bringing the system into operation. This includes, for instance,
agreement with the departmental managers as to who has ultimate
responsibility for the correctness of documents going to outside
organizations. In general, it is better for this responsibility to lie with
the user departments as they are geared up to dealing with outside
inquiries. This arrangement would not however absolve the data processing department from responsibility for the accuracy and prompti-
tude of its output.

## Instructions to user departments

It is the system analyst's responsibility to ensure that the user departments are instructed in the tasks that they have to perform in connection with the data processing system. He should also ensure that they fully understand the purpose of the system's output that they receive. These points are covered by written instructions and explanations sent by the analyst to the heads of the user departments. The points will have been explained beforehand verbally and are, in fact, a confirmation of what has already been agreed. The instructions and explanations are better expressed in advisory rather than authoritarian terms, and cover the following points:

1   The day and time by which each lot of source data is to be ready.
2   Who is responsible for preparing each lot of source documents, and to whom they are returned after being punched.
3   The preparation of input control totals and counts, exactly what these are and where they are to be shown on the source documents.
4   The precise meaning of the information on each output document.
5   Who is to receive the output documents, the number of copies, day and time of completion.
6   A brief description of file contents so that user departments' staff are aware of what information can be made available, and will not then start up their own private system to provide it.
7   What 'on request' information (if any) is available from the system, with an indication of the waiting time for it.
8   Who to contact within the data processing department in case of difficulties with the source data or output, or when changes are desired. Although this person is usually the data processing manager or the senior systems analyst, it is better to formalize the arrangement so that the user departments do not make unofficial arrangements with programmers or operators. There is, of course, no harm at a later stage in these people being involved in discussing minor changes, provided their conclusions are agreed and formally included as amendments to the system.


## Top management involvement

Top management should, to some extent, be involved in the implementation of the data processing system; there are several reasons for this:

1   Their advice and authority may be needed, for instance in order to exert pressure on managers whose departments are laggardly and endangering the implementation schedule.

2   So that they can comprehend the volume of work involved in planning and installing a data processing system, and will then give this consideration before requesting modifications to the output of the system.
3   They will acquire a feeling for data processing, and this encourages them to employ new developments in hardware and techniques in order to improve management control of the organization.
4   They will understand the reports they receive more fully and so think about their requirements for more sophisticated information. This point is important because only the managers themselves truly know their real information needs. The systems analyst must not pre-judge this issue; the demand for information should be allowed to arise as a natural consequence of top management's familiarity with the capabilities of data processing systems.

*System presentation to other staff*

This is intended to give all other staff (who are not directly concerned) a general picture of what is involved in data processing. Staff from the user departments may also attend these presentation talks if they do not feel cognizant with any aspect of the system. The talks are given by the systems analyst, possibly assisted by some of the user department staff, and aim at removing apprehension and dispelling confusion from the minds of user departments' staff.

The points to be included are:
1   A brief description of the hardware and its method of use.
2   Advantages to be gained from the new system.
3   Emphasis on human participation and responsibility.
4   Answers to any relevant questions – those relating to security of employment should be dealt with by the management rather than the systems analyst.

## 13.5   Data processing staff recruitment and training

*Recruitment of systems analysts*

It is necessary to recruit and train the data processing staff long before the computer is installed; this requirement applies especially to systems analysts and programmers. The systems analyst is involved in designing the system both before and after the decision to proceed with data processing is made, and in practice it is unlikely that he will have completed more than a small proportion of the total work when the computer is ordered. Before installation of the computer, and during the succeeding months, the remaining systems design is done, and in all but the most elementary applications the analyst continues his work after the implementation of the basic system.

In the rapidly changing field of computer-based data processing there is wide scope for steadily improving the system by introducing new hardware and techniques as these become available. It is therefore advantageous to set up a systems team on a permanent basis, its size depending upon the complexity of the organization's control and information system, and varying from one person to several scores. When building up a systems team, its structure and growth rate will be based on the advice given by the first systems analyst. He will become a key man in the company; consequently his calibre is of the utmost importance. In many cases this man is later promoted to be the data processing manager or management services manager.

In some circumstances the analyst who makes the initial recommendations is employed on a consultancy basis, coming from either an outside consultancy firm or from within the organization but not intended as a permanent systems analyst. If a team is created, there is room in it for analysts of lesser experience. These people will have the opportunity of working alongside the more experienced analysts, and can thus be trained and gain experience while working in the organization.

*Training of systems analysts*

A variety of courses are now available for the training of systems analysts, run mostly by the computer hardware manufacturers, colleges, and software and consultancy firms. Manufacturers' courses tend to be available only for their customers' staff, and a certain number of training places are sometimes offered as part of the deal. Most colleges and polytechnics run either courses based on the National Computing Centre's package or part-time courses of their own. The National Computing Centre and the British Computer Society provide lists of courses currently available.

*Recruitment of programmers*

The calibre of the programmers required depends not so much upon the complexity of the work to be programmed as on the level at which the systems analyst defines it. This level is to some extent flexible but the norm should be taken as that described in Chapter 10. It is the responsibility of the analyst to decide this level when he has had an opportunity to assess the programmers' capabilities.

Successful programmers may be recruited from a wide range of backgrounds and although a knowledge of business is a useful attribute, it is by no means essential. This means that potential programmers can be drawn from all departments within a company, including those in the factory as well as the office. Similarly they can come from a wide range of outside organizations, including other companies, or straight from school, college or university.

The essential qualities to be looked for in a potential programmer are:

1  *Education*   Up to A level standard in one or two subjects, preferably of the logical type such as physics, statistics, mathematics, chemistry and possibly languages.

2  *Aptitude for programming*   In some ways this is more important than educational qualifications, particularly if the recruit does not wish to go beyond programming in his career. There are a number of programming aptitude tests available, mostly run by the manufacturers. The results of past tests have been shown to correlate quite well with the subsequent efficiencies of the candidates as programmers [13.8].

3  *Desire to become a programmer*   It is difficult for a newcomer to data processing to comprehend what programming is about. It is therefore advisable to give applicants some idea of the nature of the work before they undergo the aptitude test and training. Because a person is successful in the aptitude test, it does not automatically follow that he or she will be content to spend several years in the occupation of computer programmer.

Whenever possible a proportion of new programming staff, say a quarter, should, when recruited, have had previous experience of writing programs in the language(s) to be used. This experience is valuable, since it not only accelerates implementation of the work to be done on the computer, but also enables raw recruits to supplement their training with advice from the more experienced staff.

*Training of programmers*

The majority of programmers are trained by the manufacturers on full-time courses lasting about four weeks. A course of this type is usually directed towards the machine (low level) language used with the manufacturer's computers. Without doubt these are the most suitable courses for programmer training; this is because they are full-time courses carried out in a computing environment and run by the originators of the relevant language.

Colleges often run part-time courses in the higher-level languages such as FORTRAN, BASIC and COBOL. These courses are suitable for gaining a general appreciation of programming, at moderate expense.

In the case of courses run by other less well-known organizations, the prospective user is advised to check their credentials most carefully before entrusting his staff to their training.

*Recruitment and training of other data processing staff*

The recruitment and training of computer operators does not normally pose a problem. They should be of O level education, and be capable of working tight schedules and following written instructions meticulously. Operators do not require a knowledge of programming, but it is probable that some of them will become programmers after a few years of operating, followed by selection and re-training. Senior operators (shift leaders) must be capable of controlling staff, and of accepting responsibility for the computer's throughput.

Operator training is usually carried out on-site by the more senior operators or initially by the manufacturer's staff.

The tape librarian and data control staff do not need to attend formal training courses. Their main attributes are the ability to organize and administer the control of files and of data entering and leaving the data processing department.

## 13.6   Implementation planning using network analysis

Network analysis is also known as the Critical Path Method or, in its more sophisticated forms, as PERT (Programme Evaluation and Review Technique). It is widely employed in industry for project planning and a large number of variations of network analysis have been developed, mostly to meet the requirements of particular industries or specialized applications. In its original form the emphasis was on the control of projects from the time aspect, but since then other aspects have become equally prominent. These include the assessment and control of resources, cost, and multi-projects; readers who require a knowledge of these aspects are referred to [13.1 and 13.4].

When employed as a technique for planning the implementation of a data processing system (the project), network analysis is mainly concerned with the time control aspect. That is, the inter-relationship of all the necessary jobs (activities), their estimated time durations, and the progressing of their achievement. Although systems implementation planning is relatively simple as compared with many other projects and, like most other projects, can be accomplished without the use of network planning, it is worthwhile employing the technique for a number of reasons:

1   This is a golden opportunity for the systems analyst and other staff to familiarize themselves with network analysis. An analyst worth his salt should be capable of utilizing it, not only for systems implementation planning but also for other projects with which he will come into contact during his career.

2   Because the network can be analysed by using the computer manufacturer's package, this provides early experience for the analyst in the use of packages and also of bureau service since this is likely to be used prior to delivery of the user's own computer [13.9].

3   Preparation of the network diagram and the estimation of activity durations forces the analyst to consider carefully all the jobs involved in system implementation, with the result that there is less chance of any activity being overlooked or under-estimated.

4   The dates within which each activity must be performed can be determined by network planning and this information used to draw up a dated schedule of activities. This facilitates the forward planning of staff requirements.

5   Calculations based on the network diagram indicate the activities that need the most attention if the scheduled dates, including the completion date, are to be maintained.

*Preparation of a network diagram*

A typical but simplified network diagram as applied to the implementation of a data processing system is shown in Figure 13.1. This diagram does not purport to cover all situations, each one must be planned on its own merits so as to include its own particular activities. Each activity is represented by an arrowed line, the length and orientation of which are of no real significance. The only rule regarding orientation is that whenever possible the arrow points in the general left to right direction as this indicates the time sense of the diagram. No attempt is made to draw the diagram to a time scale, nor is it advisable with this type of project to segregate the activities on the diagram according to departmental responsibility.

The circles separating the activities are called 'events' and represent the chronological connections between activities. An event is said to be achieved when all the activities leading up to it have been completed, and until an event is achieved no activities leading from it can be started. Thus, referring to Figure 13.1, activity 137 cannot be started until activities 133 and 136 have been completed, i.e. until event 31 has been achieved. This rule is the basis upon which the network diagram is prepared; the steps in the process are as below:

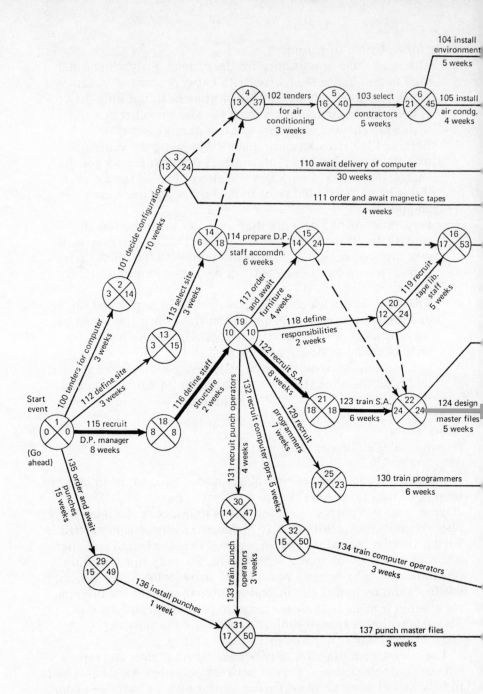

**Figure 13.1** *Network diagram for implementing a data processing system*

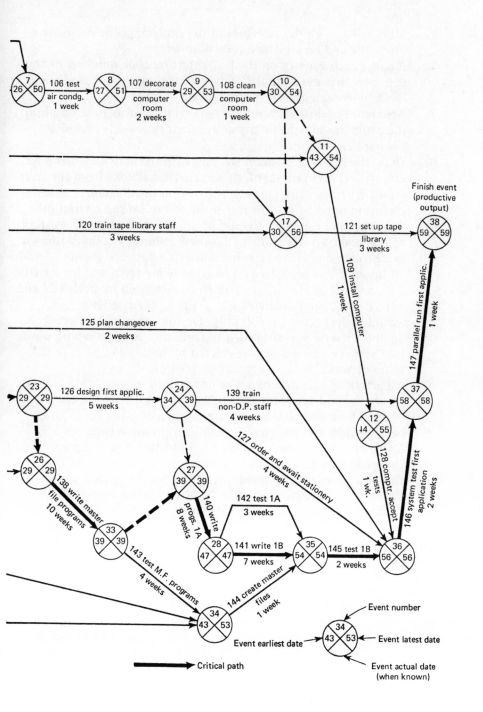

1   Make a list of all the activities in the project, giving each one a brief title and an arbitrary code number.
2   Against each activity on the list, enter the code numbers of the activities that must immediately precede it chronologically and logically (restrictions or precedences).
3   Also enter against each activity an estimate of its time duration; the estimates may be in any convenient time units provided they are consistent.
4   Draw the network diagram starting with the activities that have no restrictions upon them; these activities all lead from the start event. When positioning each activity on the diagram, particular attention must be paid to the avoidance of invalid restrictions; this is accomplished by the use of dummy activities. An example of this is shown in Figure 13.1, where a dummy activity (drawn as a broken line arrow) has been inserted between events 23 and 26 in order to stop activity 126 from being restricted by activity 130, and yet allow activity 138 to be restricted by both 124 and 130. If the dummy activity was omitted, then activity 138 would be restricted by 130 only. On the other hand if events 26 and 23 were superimposed to form one event, it would mean that activity 126 was also restricted by both 124 and 130, the latter of which is not a genuine restriction.
5   All activities that do not act as restrictions on any other activities are the end activities and these join together at the end event. These activities are apparent when drawing the diagram because they are otherwise found to form loose ends.
6   Enter each activity's code number and title on the diagram alongside its arrow. Allocate an arbitrary number to each event, and enter it in the top quadrant of the event's circle.
7   Enter the estimated time of each activity below its arrow. Dummy activities are not numbered and normally have zero time.

The network diagram is now ready for use. The calculations inherent in using the diagram involve simple arithmetic, but owing to their large volume and the intricate relationship of activities, it is easy to make a mistake. The reader is therefore advised to make use of a computer package if there are more than a hundred activities in the project. With a data processing system implementation project this is not likely to be the case, but nevertheless by utilising a package and also doing the calculations by hand, a good understanding of both the package and the technique are acquired.

*Critical path*

It is not the intention to give here a full explanation of the capabilities of network analysis — these can be found from the references. The description below covers the basic principles however.

It is obvious from even a superficial inspection of a network diagram that a large number of different paths go from its start to its finish. The project as a whole is incomplete until every activity on every path has been carried out. Each path's activities are in series which means that the minimum estimated time for a given path is the sum of the estimated times of its activities. Owing to the cross-connection of paths it is highly improbable that a path can be completed in its minimum time since its activities are delayed by other connecting activities. This implies that because one or more activities on a path are delayed, other of its activities have spare time. This argument applies to all paths except the one for which the sum of its activities' times is the largest in the network.

This is called the 'critical path' and is the one that determines the time for the project as a whole. The activities on the critical path are termed 'critical activities', and these are the ones that must be accomplished within their estimated times if the project is to be finished by the estimated date. In Figure 13.1 the critical path is shown as a heavy line, and as can be seen, it is possible for it to include several dummy activities.

*Identification of the critical path*   Bearing in mind that an event is achieved when all the activities leading up to it have been completed, it follows that the earliest achievement date for an event is that which allows for completion of the longest time path up to it. Thus in the example in Figure 13.1, consider event 15; the two paths leading up to it are

1   Activities 112, 113 and 114, taking 12 weeks.
2   Activities 115, 116 and 117, taking 14 weeks.

Since both of these paths must be completed, the earliest date for achieving event 15 is week 14.

Each event's earliest date is entered in the left-hand quadrant of its circle. The method for finding all the earliest dates is to enter nought or some other starting date at the starting event's earliest date, then to proceed forward through the network entering earliest dates whenever it is possible to do so. Each earliest date is equal to the preceding event's earliest date plus the intervening activity's estimated time; where there is a choice, the latest of the calculated earliest dates is selected. Thus, referring to event 34, there are three activities leading from three preceding events and so the set of calculations for event 34 are:

1   Earliest date of event 33 (= 39) plus estimated time of activity 143 (= 4), giving 43.
2   Earliest date of event 31 (= 17) plus estimated time of activity 137 (= 3), giving 20.

3      Earliest date of event 32 (= 15) plus estimated time of activity 134 (= 3), giving 18.

The latest of these is given by 1, so the earliest date for event 34 is week 43.

Having calculated and entered all the earliest dates, the project's estimated time is now known, i.e. the difference between the finish event's earliest date and the start event's earliest date. Because the determination of the finish event's earliest date has taken into consideration all the paths through the network, including the critical path (although this is not yet identified), this estimated date cannot be exceeded. The earliest date of the finish event is therefore the same as its latest date.

The procedure is now to work backwards through the network, entering the latest dates of the events (in the right-hand quadrants). The latest date of an event is the date until which its achievement can be delayed and yet still leave time for all its following activities to be accomplished without delaying the projector's completion date. The latest date of an event is equal to the latest date of its succeeding event minus the intervening activity's estimated time. Where there are alternatives, the earliest of the calculated latest dates is selected. For example, the calculations appertaining to event 3 are:

1      Latest date of event 4 (= 37) minus time for dummy activity (= 0), giving 37.

2      Latest date of event 11 (= 54) minus estimated time of activity 110 (= 30), giving 24.

3      Latest date of event 17 (= 56) minus estimated time of activity 111 (= 4), giving 52.

The earliest of these is given by 2, so the latest date for event 3 is week 24.

This procedure continues until the start event is reached, the latest date of which should also be calculated so as to prove that it is the same as its earliest date. This is always so, and by arriving at this conclusion logically it is probable that no mistakes have been made in the other events' dates.

It will now be observed that some events have earliest and latest dates that are the same as each other; these events are said to have zero 'slack' and they all lie on the critical path. An event's slack is equal to its latest date minus its earliest date. The critical activities are those that lie on the path through all the zero-slack events (the critical path), as can be seen in Figure 13.1.

Care should be taken not to erroneously include as critical any activities that join two zero-slack events but which by-pass others, e.g. activity 125. It is possible to have portions of parallel critical path although these are uncommon in practice; this would be the case if activity 125 had an estimated time of 32 weeks.

*Activity float*

As stated earlier, critical activities are those for which a delay in their completion causes a delay in the completion date of the overall project. Other, non-critical activities have associated with them a certain amount of spare time, known as 'float'. Within limits, the times actually taken to perform the non-critical activities can exceed their estimates without delaying the over-all project, this is because the excess times are absorbed by the floats.

The most important float is actually 'total float'. This is a measure of the absolute maximum spare time allowable to an activity on the assumption that the activity's neighbours are scheduled to cater for this. If

$A$ = activity's estimated time,
$PE$ = earliest date of its preceding event,
$PL$ = latest date of its preceding event,
$SE$ = earliest date of its succeeding event,
$SL$ = latest date of its succeeding event,

then

$$\text{Total float} = SL - PE - A$$

*Activity dates*

Each activity is surrounded by two events, each of which has an earliest date and a latest date (sometimes the same date). Thus an activity has a limit imposed on the date at which it is able to start, i.e. its earliest start date is the earliest date of its preceding event. There is also a limit on an activity's latest finish date, i.e. the latest date of its succeeding event. The non-critical activities, which have float, can be planned to take place at any time within their limits, but it must be remembered that activities' dates are mutually dependent except for those appertaining to activities with independent float.

Using the same notation as above, the four dates applicable to each activity are:

Earliest start date = $PE$
Earliest finish date = $PE + A$
Latest finish date = $SL$
Latest start date = $SL - A$

These and the total floats are given in Figure 13.2 for the network of Figure 13.1.

| Activity code | Activity title | Est. time, weeks | ES | EF | LS | LF | Total float |
|---|---|---|---|---|---|---|---|
| * 115 | Recruit D.P. manager | 8 | 0 | 8 | 0 | 8 | 0 |
| * 116 | Define D.P. staff structure | 2 | 8 | 10 | 8 | 10 | 0 |
| * 122 | Recruit systems analysts | 8 | 10 | 18 | 10 | 18 | 0 |
| * 123 | Train systems analysts | 6 | 18 | 24 | 18 | 24 | 0 |
| * 124 | Design master files | 5 | 24 | 29 | 24 | 29 | 0 |
| * 138 | Write master file programs | 10 | 29 | 39 | 29 | 39 | 0 |
| * 140 | Write batch 1A programs | 8 | 39 | 47 | 39 | 47 | 0 |
| * 141 | Write batch 1B programs | 7 | 47 | 54 | 47 | 54 | 0 |
| +* 145 | Test batch 1B programs | 2 | 54 | 56 | 54 | 56 | 0 |
| * 146 | System test first application | 2 | 56 | 58 | 56 | 58 | 0 |
| * 147 | Parallel run first application | 1 | 58 | 59 | 58 | 59 | 0 |
| + 142 | Test batch 1A programs | 3 | 47 | 50 | 51 | 54 | 4 |
| 126 | Design first application | 5 | 29 | 34 | 34 | 39 | 5 |
| 129 | Recruit programmers | 7 | 10 | 17 | 16 | 23 | 6 |
| 130 | Train programmers | 6 | 17 | 23 | 23 | 29 | 6 |
| 117 | Order & await furniture | 4 | 10 | 14 | 20 | 24 | 10 |
| + 143 | Test master file programs | 4 | 39 | 43 | 49 | 53 | 10 |
| + 144 | Create master files | 1 | 43 | 44 | 53 | 54 | 10 |
| 100 | Accept computer tenders | 3 | 0 | 3 | 11 | 14 | 11 |
| 101 | Decide configuration | 10 | 3 | 13 | 14 | 24 | 11 |
| 110 | Await delivery of computer | 30 | 13 | 43 | 24 | 54 | 11 |
| 109 | Install computer | 1 | 43 | 44 | 54 | 55 | 11 |
| 128 | Computer acceptance tests | 1 | 44 | 45 | 55 | 56 | 11 |
| 112 | Define computer site | 3 | 0 | 3 | 12 | 15 | 12 |
| 113 | Select computer site | 3 | 3 | 6 | 15 | 18 | 12 |
| 114 | Prepare D.P. staff accomdn. | 6 | 6 | 12 | 18 | 24 | 12 |
| 118 | Define staff responsibilities | 2 | 10 | 12 | 22 | 24 | 12 |
| 127 | Order & await stationery | 4 | 34 | 38 | 52 | 56 | 18 |
| 139 | Train non-D.P. staff | 4 | 34 | 38 | 54 | 58 | 20 |
| 102 | Accept air condg. tenders | 3 | 13 | 16 | 37 | 40 | 24 |
| 103 | Select air condg. contrs. | 5 | 16 | 21 | 40 | 45 | 24 |
| 104 | Install environment | 5 | 21 | 26 | 45 | 50 | 24 |
| 106 | Test air condg. plant | 1 | 26 | 27 | 50 | 51 | 24 |
| 107 | Decorate computer room | 2 | 27 | 29 | 51 | 53 | 24 |
| 108 | Clean computer room | 1 | 29 | 30 | 53 | 54 | 24 |
| 105 | Install air condg. plant | 4 | 21 | 25 | 46 | 50 | 25 |
| 121 | Set up tape library | 3 | 30 | 33 | 56 | 59 | 26 |
| 125 | Plan changeover | 2 | 24 | 26 | 54 | 56 | 30 |
| 131 | Recruit punch operators | 4 | 10 | 14 | 43 | 47 | 33 |
| 133 | Train punch operators | 3 | 14 | 17 | 47 | 50 | 33 |
| 137 | Punch master files | 3 | 17 | 20 | 50 | 53 | 33 |
| 135 | Order & await punches | 15 | 0 | 15 | 34 | 49 | 34 |
| 136 | Install punches | 1 | 15 | 16 | 49 | 50 | 34 |
| 132 | Recruit computer operators | 5 | 10 | 15 | 45 | 50 | 35 |
| 134 | Train computer operators | 3 | 15 | 18 | 50 | 53 | 35 |
| 119 | Recruit tape library staff | 5 | 12 | 17 | 48 | 53 | 36 |
| 120 | Train tape library staff | 3 | 17 | 20 | 53 | 56 | 36 |
| 111 | Order & await mag. tapes | 4 | 13 | 17 | 52 | 56 | 39 |

NOTES

* Critical path (shown as heavy lines in Figure 13.1)

+ On bureau computer prior to installation of user's computer.

**Figure 13.2** *Activity dates and total floats for Figure 13.1*

Network planning is not intended to apply merely to static situations. The initial network having been drawn, and the activity dates and floats calculated, these are then used as a means of keeping a check on the progress of the project throughout its life. When activities are completed – completely or partially, the network is re-analysed so that critical and near critical activities are detected and can be closely watched. It is, of course, quite possible that the critical path will change its course several times during the life of a project. This is brought about by the insertion and deletion of activities, by amendments to estimated times, and by the actual dates of completion of activities.

If a computer package is employed to analyse the network, the current position of the project is held on magnetic tape or disk, and updated at regular intervals when amendments occur. It is not necessary to re-draw the network diagram at any time unless very drastic changes are made to it. The outputs of packages are very flexible, the information being producible in various sequences so as to emphasize the main points of interest in a particular project. In addition to the conventional listed output such as in Figure 13.2, the activity dates and floats can be printed by the computer in the form of bar charts. The list in Figure 13.2 is in earliest start within total float sequence; this means that the most critical activities appear at the beginning of the list and are thereby most easily recognized by the project controller.

When using network analysis for system implementation, a careful inspection of the progress reports enables the systems analyst to anticipate difficulties in relation to the completion dates of the system implementation activities. This means that measures can be adopted to avoid hold-ups, and thereby ensure that the data processing system will commence operations on time.

## 13.7   System monitoring

The purpose of system monitoring is to keep a watch on the efficiencies of the data processing routines by making regular checks after their implementation. Although a properly designed data processing system always produces complete and accurate output results, it is possible that changing circumstances will prevent it from doing so in an efficient manner. At the time of designing the system it is not always possible to foresee the changes that can come about, and subsequent system monitoring is therefore advisable in order to detect and counteract these changes.

The two main factors with which the systems analyst is concerned during monitoring are the times taken by the routines and the utilization of storage. These factors are tied together closely in that inefficient use of serial or direct access storage may cause an increase in the run times.

Every major routine should be monitored at regular intervals, say six-monthly, by the systems analyst or another nominated person. This involves inspecting operating system reports and/or console log sheets in order to correlate data volumes with run times. The necessary facts to be determined for each run are those such as:

The number of items input and/or output.

The number of lines printed.

The activity or volatility of the files.

The time taken for the run, both processing time and overall time.

By examining the above figures in conjunction with his knowledge of the routines, it is feasible for the systems analyst to detect inefficiencies in the system. These may have been introduced through a number of causes, principal among which are:

1   Changes in the sets of entities in use resulting in much 'dead wood' being left in the files, e.g. disused parts, obsolete products.
2   An overloading of a direct access file giving rise to an excess of overflow records.
3   The introduction of new entities with unsuitable code numbers, causing the address generation algorithms to become less efficient and thereby lowering the uniformity of record distribution in a random file.
4   Omissions of output results caused by deliberate or unintentional operating errors.


*Contact with the other departments*

Another aspect of system monitoring is the maintenance of contact between the systems analyst and the recipients of the system's output. It is essential that the objectives and output of the system are kept up to date, and not allowed to become obsolete through default. The degree to which contact is maintained depends upon the type of organization and the applications involved. In some cases there is inherently no change in an application for many years, in others the ability of the system to change course quite frequently is of great value. Although a data processing system should be designed to accept a fair degree of foreseeable changes, there may be circumstances that arise unpredictably and therefore call for special arrangements. These are discovered as a consequence of maintaining contact with the user departments, and asking the staff therein questions such as:

What further requirements or changes are needed in future?

Are any of the output results now redundant?
Are there any impending changes of consequence to the sets of entities?
As far as is known, will the input volumes increase significantly in the near future?

## 13.8   Computer manufacturers' services

Following the signing of the contract for the purchase or rental of a computer, the user may have to rely heavily upon support by the manufacturer. That is especially true if the user has no previous experience of data processing. It is therefore important that the decision as to which manufacturer to order from is based not only on the characteristics and cost of hardware, but also upon the services that the manufacturer can offer. Having realized this, the user should find out from each prospective supplier what services are available, and at what cost, but at the same time endeavour to become self-supporting as soon as possible. A policy of leaning on the manufacturer and obtaining every iota of service — based on the philosophy of getting one's money's worth — is both risky and inefficient. The inevitable withdrawal sooner or later of this support invariably exposes a creaking data processing department.

What services can a user expect to receive, and to what extent are these covered by the contract? The latter point depends on what special arrangements, if any, are agreed between the manufacturer and the user. Generally it is only computer maintenance that is covered by the terms of the standard contract, and this is so because an additional charge is made for it — although maintenance charges may be included in the rental. The other services described below are obtainable to different extents, depending on *(a)* the manufacturer's ability to provide them and *(b)* the user's ability to force the manufacturer to provide them, and *(c)* the manufacturer's unbundling policy, i.e. separation of software charges from hardware charges.

*Training services*

The computer manufacturers all have training schools, and generally speaking, these are the best available, particularly for lower level courses. The prospective user should obtain the manufacturers' training course prospectuses, and from them discover the range, cost and frequency of the courses held. The full spectrum of training includes courses required by top management, middle management, data processing management, systems analysts, programmers (elementary and advanced), and the various operators. By and large these courses are held at the manufacturer's residential schools on a full-time basis, the period involved varying from one or two days for top management courses, up to a few weeks for programming courses. Exceptions to this arrangement are courses that are run specially for one customer on his own premises, also operator training, which is usually also carried out locally. Self-tuition kits based on teaching machines and special literature are also available sometimes.

The over-all cost of training staff at all levels can amount to a considerable sum, but set against this are the 'free' or reduced price courses that may be available to new users. The value of training discounts is normally based upon the value of the computer ordered.

*Programming support*

This takes two forms, firstly actual program writing done for the user by the manufacturer's programmers. However, unless the value of the computer order is very large, it is unlikely that any free support of this type will be forthcoming. Nevertheless there are chargeable programming services available from programming service firms (software houses). The biggest problem in utilizing these services is communication, but provided the systems analyst fully specifies the runs (as described in Section 10.8), then the service should be effective. These services tend to be expensive but are useful for helping out with overloads of programming such as might be experienced in the early stages of implementation. It would be unwise of a computer user to employ a programming service to the complete exclusion of having his own programmers.

The second form of programming support is advisory, the manufacturer's programmers being available to advise the users' programmers during the first few months after the latter's initial training. The need for this support diminishes rapidly but it can be extremely valuable if the user has no experienced programmers of his own.

*Specialized advice*

This is connected with the design of special systems and the employment of sophisticated techniques. These include the various operational research techniques, network planning and PERT, production control and other application packages, and knowledge of the user's industry or organization. Although a manufacturer cannot be expected to have a comprehensive team of specialists in every local branch office, these people should be available from elsewhere at reasonably short notice and be prepared to make occasional visits to the user. If the user's systems analyst feels that specialized advice is vital to the system, its requirement should be written into the contractual agreement with the manufacturer. As in the case of programming advice, the specialist must be of the right calibre with practical experience of implementing the application in which he specializes.

## Program testing

Prior to delivery of the user's computer, it is essential that sufficient programs have been written and tested to enable the computer to produce useful results immediately after it is installed. To have it standing idle, apart from program testing, for perhaps several months, is both uneconomic for the company and damaging to the data processing department's prestige. Facilities must therefore be provided for program testing on another computer, preferably of a similar configuration to the user's own computer. As with training, program testing quickly accumulates a substantial cost, and it is therefore wise to obtain a written statement from the manufacturer specifying the facilities available and their cost.

## Installation advice

The prospective user, having acquired sufficient knowledge about computers to venture placing an order, then finds himself thrust into the world of air conditioning and environment preparation. As with data processing itself and most other complex subjects, the first problem is to understand the jargon. Unfortunately time does not allow for this, with the result that the user is soon confounded by the cross-claims and varying specifications of the suppliers of air conditioning plant and environment equipment.

The computer manufacturer is ethically obliged to render assistance by making available his environment specialist to discuss the problems with the air conditioning suppliers' representatives. He should also advise on the form of tender and the best quotation for the user to accept.

## Computer maintenance

The computer, in common with other complex equipment, requires repair and regular maintenance; these apply particularly to its mechanical parts. The user requires the manufacturer to state how much time is required for scheduled maintenance, and to arrange with him the times at which this will be carried out. Other factors relating to maintenance that are significant are:

231

1    The manufacturer's facilities for coping with machine faults and breakdown. If two or more manufacturers are involved in supplying interconnected equipment, the maintenance interface should, as far as is possible, be established from the start. This applies to data transmission equipment, for which the responsibility for faults can so easily be tossed back and forth between the suppliers.
2    Where are the maintenance engineers based, and what is the ratio of computers to engineers in the area? If the user's computer is large, it is not unreasonable to expect to have a resident maintenance engineer.
3    What reservoir of maintenance service can be called in if there is a major breakdown or persistent trouble?
4    What is the situation regarding spare parts, where are these held, and what is the maximum time taken to obtain any spares, including major components?

### Stand-by arrangements

The two main reasons for being interested in stand-by arrangements are *(a)* in case of persistent breakdown and *(b)* to cope with unforeseen temporary overloads. The latter tends to occur as a consequence of the former.

The most convenient stand-by arrangement is to arrive at a *quid pro quo* understanding with another local user who has a similar configuration. The more informal and the less financial this is, the better. If his configuration is usable but not identical, it is worth preparing modified versions of the important programs to suit his configuration.

Failing the above arrangement, the manufacturer's computers are the next best bet. The questions to be answered here are:

What configurations are available and where are they situated?
Are these computers available solely for stand-by purposes?
What financial arrangements are involved?

Computers installed by the manufacturer for software development or service bureau work are generally too heavily committed to be suitable for stand-by purposes.

### Service bureau facilities

The reasons for employing bureau service were expounded in Section 12.3; it is also common for users to employ bureaux for master file preparation (Section 13.2) during the system implementation phase.

The factors to be investigated in relation to proposed bureau service are:

1    Who owns the programs written by the bureau's programmers but paid for by the user? It does not always follow that the payment of perhaps several hundred pounds to have a program prepared, entitles the user to have a copy of it for either using on his own computer or re-selling.

2    The quotations received from several bureaux, based on identical job specifications, turn-round times, etc., should be compared. These may vary, in regard to programming charges, file creation and job set-up charges, and regular running charges.

3    The quality of the bureau's service should be discussed with other users of the bureau. This applies to turn-round time, accuracy, completeness and presentation of output, and general reliability. Also of interest is the continuity of employment of the bureau's staff; if this is short-lived, it is difficult for the user to build up his bureau routines because the bureau staff never become really familiar with his problems.

4    As with all contractual agreements, the prospective bureau user should read the small print on the back of the contract form most carefully.

Membership of the Computing Services Association (CSA) indicates that the bureau company is prepared to accept certain standards. These are available from the CSA at 109 Kingsway, London WC2B 6PU. Similar associations exist in other countries.

## 13.9 References and further reading

13.1    WOODGATE, *Planning by network*, Business Books (1977).

13.2    LOCKYER, *Introduction to critical path analysis*, Pitman (1968).

13.3    WARD, *Computer organization, personnel and control*, Longman (1973).

13.4    ROBERTSON, *Project planning and control*, Heywood (1969).

13.5    'Getting your man', *Data Systems* (May 1970).

13.6    'Implications of conversion', *Data Processing* (January-February 1971).

13.7    'Computer services and the law', *Business Automation* (November 1970).

13.8    'New look at programming aptitudes', *ibid.* (August 1970).

13.9    *PERT users' guide*, I.C.L. technical publications 4200, 4133 and 4059.

13.10  CLIFTON, *Choosing and using computers*, Business Books (1975).

# Appendix

# *Example of a computer run specification*

This example is the specification of run K12 in Figure 10.5, and is intended for a programmer who is reasonably experienced and familiar with the data processing department's standards.

*General description of run K12*

This run reads the invoice file K1 in product code sequence, matches it against the product master cost file T2, also in product code sequence, and creates a new file K2 (costed invoice file). At the same time, unmatched product codes from K1, discrepancies and control totals are printed.

*Volumes of data*

File K1 — 5000 to 7000 records per weekly run.
File K2 — 5000 to 7000 records per weekly run.
The above two files each have one record per invoice item, unmatched records being omitted from K2. There may be any number of records (including zero) for a given product code.
File T2 — 600 records approximately, one per product.
Document K22 — 100 print lines approximately.

## Layouts of files

### File K1 — Invoice items

| Data item | Picture | Positions |
|---|---|---|
| Product code | 99999 | 1–5 |
| Invoice number | A9999 | 6–10 |
| Area | 99 | 11–12 |
| Class of trade | 99 | 13–14 |
| Customer number | 999 | 15–17 |
| Quantity | 9999 | 18–21 |
| Product value | 999.99 | 22–26 |
| Discount value | 999.99 | 27–31 |
| Week number | 99 | 32–33 |
| Year | 99 | 34–45 |

### File T2 — Product Cost Master

| Data item | Picture | Positions |
|---|---|---|
| Product code | 99999 | 1–5 |
| Standard labour price | 99.9999 | 6–11 |
| Standard material price | 99.9999 | 12–17 |
| Maximum discount, per cent | 99.999 | 18–22 |
| Selling price | 9.99 | 23–25 |
| Description | A20 | 26–45 |

### File K2 — Costed invoice items

| Data item | Picture | Positions |
|---|---|---|
| Area | 99 | 1–2 |
| Class of trade | 99 | 3–4 |
| Customer number | 999 | 5–7 |
| Product code | 99999 | 8–12 |
| Standard labour value | 999.9999 | 13–19 |
| Standard material value | 999.9999 | 20–26 |
| Product value | 999.99 | 27–31 |
| Calculated value | 999.99 | 32–36 |
| Discount value | 999.99 | 37–41 |
| Week number | 99 | 42–43 |
| Year | 99 | 44–45 |
| Quantity | 9999 | 46–49 |

## BUCKET LENGTHS

File K1 1000 characters = 28 records
File K2 1000 characters = 20 records
File T2 2000 characters = 44 records

## Processing

1   Check week and year as per parameters, if not stop run, display as per operating instructions.
2   Read K1 record and match its product code against T2 — if matched, add 1 to control count H — if unmatched, add 1 to control count J, print as per output layout line A, and ignore K1 record.
3   Extract selling price from T2, multiply by quantity equals calculated value.
4   Compare calculated value with product value from K1, if un-

equal – add 1 to control count K, and print as per output layout line B.

5    Add calculated value to control total D.
Add product value to control total E.
Add quantity to control total G.

6    Extract standard costs from T2, multiply by quantity, and insert results into K2.
Standard labour value = standard labour price × quantity.
Standard material value = standard material price × quantity.
No rounding is necessary.

7    Multiply calculated value × maximum discount per cent = maximum discount value.

8    Compare maximum discount value with discount value from K1. If discount value is greater than maximum, add 1 to control count L, and print as per output layout line C.

9    At end of K1, calculate total value discrepancy = total calculated value less total product value, and print as per output line F, with * if positive, – if negative.

## Output layouts

Since this is an internal document to be printed on blank stationery, the exact print positions are not important but the layout is as shown in the example below.

Number of copies: 2
Spacing: double
Lines per sheet: 30
Sheets numbered consecutively.

| | Product Code | Invoice No. | Calculated Value | Product Value | Maximum Discount | Discount Value | Comment |
|---|---|---|---|---|---|---|---|
| A | 44697 | E8093 | | | | | Invalid produce code |
| B | 48560 | E8562 | 73.62 | 72.52 | | | Value discrepancy |
| C | 51233 | E8411 | | | 15.56 | 23.34 | Discount above maximum |
| D | Total calculated value | | | 73826.57 | | | |
| E | Total product value | | | 73596.51 | | | |
| F | Total value discrepancy | | | 230.06* | | | |
| G | Total quantity | | | 131265 | | | |
| H | Matched records count | | | 6318 | | | |
| J | Unmatched records count | | | 46 | | | |
| K | Value discrepancy count | | | 28 | | | |
| L | Discount above maximum cost | | | 16 | | | |

# *Index*

238